Water Resources Development and Management
Series Editors: Asit K. Biswas and Cecilia Tortajada

Chennat Gopalakrishnan
Cecilia Tortajada
Asit K. Biswas
Editors

Water Institutions: Policies, Performance and Prospects

with 6 Figures and 12 Tables

THIRD WORLD CENTRE FOR WATER MANAGEMENT

Springer

Professor Chennat Gopalakrishnan
University of Hawaii at Manoa
College of Tropical Agriculture
and Human Resources
East-West Road 1910
Sherman Lab 118
Honolulu, HI 96822-2279
USA

Dr. Cecilia Tortajada
Third World Centre
for Water Management
Avenida Manantial Oriente 27
Los Clubes, Atizapan
Estado de México 52958
Mexico

Professor Asit K. Biswas
Third World Centre
for Water Management
Avenida Manantial Oriente 27
Los Clubes, Atizapan
Estado de México 52958
Mexico
Email: akbiswas@att.net.mx

Cover photo: © Haroldo Palo, Jr.

Library of Congress Control Number: 2004115724

ISSN 1614-810X
ISBN 3-540-23811-5 Springer Berlin Heidelberg New York

Springer is a part of Springer Science+Business Media
Springeronline.com

Printed in Germany

Typesetting: Camera ready by authors
Cover design: E. Kirchner, Heidelberg
Production: Almas Schimmel
Printing: Mercedes Druck, Berlin
Binding: Stein + Lehmann, Berlin

Printed on acid-free paper 30/3141/as 5 4 3 2 1 0

Preface

It is being increasingly realised that water is likely to be one of the most critical resource issues for the first half of the twenty-first century. Accelerating demand for water for various uses and user groups and ineffective measures to address water quality decline from point and non-point sources of pollution, have made water management more complex and difficult than ever before in human history. All the current trends indicate that water management will become even more complex in the future because of society's higher demands for good quality water, and new and emerging impacts on the water sector due to the forces of globalisation. These include the liberalisation of trade in agricultural and manufactured products, information and communication revolution, and technological developments in areas traditionally not considered to be water-oriented, like biotechnology. Impacts of these new and emerging forces on the water sector are still not fully understood or appreciated at present, but they are likely to change water use practices dramatically in many countries of the world during the coming decades.

While it is now generally accepted that water management in terms of quantity and quality will be a very difficult and complex task in the coming years because of accelerating human activities, it is a curious anomaly that studies and analyses of institutions (both public and private) which manage water, have been a most neglected subject in the past. This is difficult to understand, since in the final analysis how efficiently and equitably water is managed in different parts of the world primarily depends on the capacities of the institutions that plan and manage them.

If water management is to become efficient globally, nationally and sub-nationally, there is no question that the institutions that manage this resource must become increasingly more and more competent. Unless water management institutions become more efficient, improvements in the governance of water are likely to be at best slow and incremental. As the World Commission on Water for the 21st Century, of which one of the Editors was a member, has noted in its recent (2001) report: "… … with current institutional arrangements and current technologies, the arithmetic of water doesn't add up." The Commission then goes on to say that only rapid and imaginative institutional and technological innovations can avoid a water crisis.

Because of the facts that institutions play a most critical role in water management, and that objective studies and analyses of water institutions have been a most neglected subject in the past, it was decided to prepare a book which addresses comprehensively and authoritatively some of the fundamental aspects of institutions that manage water from different parts of the world.

The term "institutions" has often been defined differently by different authors. In the present context, we have taken a broad approach toward institutions by incorporating the key elements identified in seminal writings on institutions that span several decades. For the purposes of this volume, institutions are viewed as having three essential components: water organizations and agencies at the local, state, and national levels entrusted with the delivery, allocation, transfer, and management of water among uses and users; the laws, rules, regulations, ordinances, etc. that govern the ownership and allocation of water; and the cultural, political, and technological settings in which the water organizations are embedded and through which the water laws are shaped.

This book consists of nine original papers by a team of scholars and experts in water institutions and policies. The first chapter by Gopalakrishnan presents a review and discussion of the concept of institutions and the idea of institutional entropy by appraising Hawaii's water institutions. It sets the stage for the discussions and analyses in the chapters that follow. The next three chapters by Biswas, Saleth, and Nickum explore the central issues pertaining to water institutions in three Asian countries of Sri Lanka, India, and China, respectively, and offer suggestions for institutional reforms and strengthening. Chapter 5 by Tortajada and Contreras-Moreno provides a comprehensive analysis of the evolution of river basin institutions in Mexico, their restructuring and overall performances during the past half century. The next chapter by Beaumont gives an evolutionary perspective and analysis of water institutions in the Middle East. Post-apartheid institutional developments in the South African river basins are reviewed and analysed by Turton and Earle in Chapter 7. The final two chapters deal with water institutions in the Western United States. In Chapter 8, Howe explores property rights and water rights in the context of the changing configuration of water allocation and use. Chapter 9 by Huffaker is an analysis and critique of the role of the Doctrine of Prior Appropriation in meeting the changing and challenging water needs of the Western United States.

In essence, institutional innovations could play a decisive part in improving water management practices and processes in the world, and also in improving the human welfare and quality of life in the developing world. The literature on the role of institutional innovations and change as they pertain to natural resources in general, and water in particular, is sparse. This volume is intended to address this gap. The papers selected for inclusion in this book examine both the conceptual and empirical dimensions of institutional innovations, through the design and implementation of sustainable institutions for water resources planning, development and management. The findings from these analyses should have a good measure of applicability to institutional design for the management of other natural resources, as well, in local, regional, and international settings and scales.

Most of the issues discussed in these pages will continue to be of concern to developing countries as well as economically advanced countries during this decade and beyond. As such, comprehensive and in-depth investigations and analyses of many aspects only briefly touched on in this book are clearly warranted. We hope that this volume will spark the interest of the water professionals and others to undertake further research on the institutional dimensions of water resources management, a subject that has been conspicuous by its absence thus far. Studies of water institutions, and especially how they should respond successfully to the changing global, regional and national scenes, are urgently needed.

This volume is primarily aimed at a wide-ranging constituency of water professionals. This broad group includes hydrologists, economists, policy analysts, planners, administrators, lawyers, historians, governmental and non-governmental entities, international agencies, environmental groups, funding agencies, and academics and students from water-related disciplines.

The editors would like to express their special appreciation to Ms. Thania Gómez of the Third World Centre for Water Management for putting the entire manuscript in the format required by the publisher.

Chennat Gopalakrishnan
Cecilia Tortajada
Asit K. Biswas

Contents

List of Contributors

Beaumont, Peter, Professor, Geography Centre for Environmental Management, University of Wales Lampeter Llanbedr Pont Steffan, Ceredigion, SA48 7ED,Wales, UK

Biswas, Asit K., President, Third World Centre for Water Management, Avenida Manantial Oriente No. 27, Los Clubes, Atizapán, Estado de México, 52958, Mexico

Contreras-Moreno, Nancy, Advisor, Inter-American Institute for Cooperation on Agriculture, Organization of American States, Insurgentes Sur 1106-5, Col. del Valle, 03100, DF, Mexico

Earle, Anton, Deputy Head, African Water Issues Research Unit (AWIRU), University of Pretoria, 0002 Pretoria, Republic of South Africa

Gopalakrishnan, Chennat, Professor, Department of Natural Resources and Environmental Management, 1910 East-West Road, Sherman Lab 118, University of Hawaii at Manoa, Honolulu, Hawaii 96822-2279, USA

Howe, Charles W., Professor Emeritus of Economics and Senior Scholar, Institute of Behavioral Science, University of Colorado-Boulder, CO 80309-0468, USA

Huffaker, Ray, Professor, Department of Agricultural and Resource Economics, Washington State University, Pullman, WA 99164, USA

Nickum, James E., Professor, Tokyo Jogakkan College, 1105 Tsuruma, Machida-shi, Tokyo 194-0004, Japan

Saleth, Maria R., Senior Institutional Economist, International Water Management Institute, PO Box: 2075, Colombo, Sri Lanka

Tortajada, Cecilia, Vice President, Third World Centre for Water Management, Avenida Manantial Oriente No. 27, Los Clubes, Atizapán, Estado de México, 52958, Mexico

Turton, Anthony R., GIBB-SERA Chair in Integrated Water Resource Management, Building 21, Environmentek, CSIR, P.O. Box 395, Pretoria, 0001, Republic of South Africa

Water Allocation and Management in Hawaii: A Case of Institutional Entropy

Chennat Gopalakrishnan

1.1 Introduction

Recent years have seen a sharp increase in the demand for water in Hawaii. Many factors account for this growth. These include population growth, expansion in the visitor industry, and increasing water demands in the environmental, urban, and industrial sectors, among others. The increase in water demand has intensified and accelerated the competition for the state's water resources. This rapidly changing configuration of water demand has resulted in a number of conflicts in the allocation of water among its competing uses and users, largely because of the failure of Hawaii's water institutions, which are responsible for the planning, allocation, and management of the state's water resources. This chapter is a first attempt to study the evolution, structure, performance, and prospects of Hawaii's water institutions in terms of the concept of "institutional entropy" (see 1.3 below). The term "water institutions", for the purposes of this paper, is defined as consisting of three elements: water laws; political processes; and water administration.

Following the Introduction (Part 1.1), Part 1.2 gives an overview of important recent discourse on the notion of institutions. The concept of institutional entropy is defined and discussed in part 1.3. Part 1.4 describes and evaluates the three components of Hawaii's water institutions – water laws, political processes, and water administration. Then, Part 1.5 reviews and critically evaluates the performance of water institutions in Hawaii today, using a specific case study. Drawing on the study findings, Part 1.6, provides an original critique of Hawaii's water institutions using the concept of institutional entropy. The future directions for Hawaii's water institutions are set forth in Part 1.7.

1.2 Institutions: An Overview

There is no single, universally accepted definition of the term "institutions". A survey of the literature on this topic shows a pronounced diversity and range of thinking on the concept of institutions. Nevertheless, a careful scrutiny will reveal a few common themes. In the rest of this section, I present a sampling of this discourse.

Commons (1968) conceives of institutions in terms of "What the individuals must or must not do (compulsion or duty), or what they may do without interference from other individuals (permission or liberty), what they can do with the aid of the collective power (capacity or right), and what they cannot expect the collective power to do in their behalf (incapacity or exposure)."

Ruttan and Hayami (1984) view "institutions (as) the rules of a society or of organizations that facilitate coordination among people by helping them form expectations which each person can reasonably hold in dealing with others."

According to North (1989), "Institutions are rules, enforcement characteristics of rules, and norms of behaviour that structure repeated human interaction." He defines rules as "Constitutions, statutes and common laws, and contracts (which) specify in formal terms the rules of the game, from the most general constitutional ones to the specific terms of a particular exchange." Norms are defined as "informal constraints on behaviour that are in part derivative of formal rules; that is, they are extensions of such rules and apply to specific issues." North goes on to point out that norms, more importantly, are "codes of conduct, taboos, standards of behaviour that are in part derived from perceptions that all individuals form both to explain and to evaluate the world around them. Some of these perceptions are shaped and moulded by organized ideologies (religions, social and political values, etc.). Others are honed by experience, which leads to the reaffirmation or rejection of earlier norms."

Ostrom et al. (1993) consider institutions as people and the patterns of regular repetitive interactions among them that transform inputs into outputs. They offer as examples of institutions, families, private firms, government agencies, and local communities, among others.

Pejovich (1995) defines institutions "as the legal, administrative, and customary arrangements for repeated human interactions. Their major function is to enhance the predictability of human behaviour".

Aoki (2001) states that "An institution is a self-sustaining system of shared beliefs about how the game is played. Its substance is a compressed representation of the salient, invariant failures of an equilibrium path, perceived by almost all the agents in the domain as relevant to their own strategic choices. As such, it governs the strategic interactions of agents in a self-enforcing manner, and in turn, reproduced by their actual choices in a continually exchanging environment."

Nabli and Nugent (1989) have identified the following characteristics of institutions: 1) organizational content i.e. the extent to which organizations and institutions coincide; 2) degree of formal or informal arrangement; 3) specificity of time, place and means; 4) embeddedness in other institutions; 5) universality of the interests they serve; 6) impact on public good; and 7) link to technology.

Aoki (2001) identifies five characteristics that are implicit in this conceptualization: endogenicity, information compression or summary representation, robustness or durability with respect to continual environmental changes and agents' minor deviance from the implied rules, universality of relevance to all agents in a domain, and multiplicity.

In the analysis that follows, we will review and critique Hawaii's water institutions as they have evolved over time using some of the ideas presented here. Clearly, institutions are the products of the societies in which they are embedded, and thus may have unique characteristics and dimensions that do not necessarily come under or fit into the typology noted above. This fact will be given due attention in this analysis.

1.3 Institutional Entropy

The concept of entropy as embodied in the second law of thermodynamics tells us that matter undergoes an incremental diminution in quality with each successive use (Greven, Keller and Warnacke 2003). This concept captures a truth not necessarily limited to physical phenomena, but also has compelling relevance to social institutions (Guiasu and Guiasu 2003; Karmeshu 2003; Rifkin 1989). In the context of institutions, the idea of entropy refers to the progressive decrease in effectiveness and efficiency in performing the goals and objectives as originally envisioned and set forth. I argue in this paper that this "dysfunctionality" of institutions is largely attributable to the intrusion of entropy, which causes disarray in the inner workings of the affected institutions and thus renders them diminished in their ability to perform at peak efficiency.

In this chapter, I attempt to identify the constraints that might lead to institutional entropy. This is a preliminary profile and is not intended as a complete listing of all possible contributory factors. First, as changes in politics, economics, technology, and lifestyle occur in a society, the institutions which are embedded in it, must have the flexibility or adaptability to cope with these changes. Absent such flexibility, the ability for effective performance will be compromised. Second, a key feature or attribute of an effective institution is autonomy. The institution should be free from internal as well as external pressures to manipulate water policy. Lack of autonomy, thus, will force institutions to stray from the optimal decision-making path and lead to undesirable outcomes. Third, full and free access to all pertinent information and data that may have a bearing on making efficient and equitable decisions must be an indispensable attribute of optimal institutions. In terms of both quality and quantity, the empirical and policy data should be reliable, verifiable, and timely. Lack of access to such information inevitably leads to the onset of institutional entropy. A fourth feature of institutional entropy is a gradual erosion in what I would call "cultural calibration". By this term, I mean that the institutions must be firmly anchored in the local cultural milieu in order to effectively capture and absorb the aspirations, preferences, and unique sensibilities of the multiple stakeholders in a given community, of which the institution is an integral part.

In many countries, resource management institutions initially designed to perform at top efficiency have failed to make adequate and appropriate adjustments to allow for the many changes – political, technological, legal, and cultural – that inevitably accompany the passage of time. As a result, there is, I hypothesize, an incremental accumulation of entropy that would render the institutions progressively dysfunctional. In some instances, the institutions could become altogether obsolete and thus outlive their utility.

Examples of institutions rendered partially or totally ineffective abound in the literature on institutions. The Doctrine of Prior Appropriation, the legal doctrine governing water ownership, allocation, and management in the western United States, is a compelling case in point (Huffaker et al. 2000; Gardner 2003; Howe 2004). Other examples include water institutions in India (Vaidyanathan 1993; Saleth 2004), Sri Lanka (Gunatilake and Gopalakrishnan 2002), Middle East (Beaumont 2004), and China (Nickum 2004), among others. Later in this chapter, using the criteria noted above, I will evaluate Hawaii's water institutions and demonstrate how they have been severely constrained by institutional entropy.

1.4 Hawaii's Water Institutions: An Evolutionary Perspective

In this section, I provide a historical perspective of the origins, evolution, and development of Hawaii's water institutions in terms of their three components: 1) water laws, customs, and traditions from the ancient Hawaiian period to the present; 2) the political processes that have influenced and shaped the rules and regulations governing water ownership, allocation, and management; and 3) state, local, and national agencies, as well as private groups, that have played formal and informal roles in the context of water administration.

1.4.1 Water Laws: Formal and Informal

The ancient Hawaiian system of water rights was unique in that it acknowledged water as a public good, a natural bounty that belonged to the people, and accorded the rulers the role of custodians entrusted with the task of managing it effectively and equitably. The Hawaiian word for water is "wai" and the term for wealth is "wai wai", thus signifying the importance given by the ancient Hawaiians to water by equating water with wealth.

Watershed management in Hawaii can be traced to the original settlers of the islands. "Many scholars believed that the first inhabitants arrived in Hawaii from the Marquesas Islands between 300 and 600 A.D., although Hawaiian oral tradition indicates it may have been as early as the 1st century A.D. Archaeological evidence suggests that the early migrants settled along the coast near fresh water resources, primarily in the windward valleys, and practiced a mixture of shifting cultivation agriculture and subsistence fishing (Kirch 1985)." (Derrickson et al. 2002)

The present system of water laws in Hawaii, for analytical purposes, can be described in terms of four periods: pre-contact, pre-McBryde, post-McBryde, and Water Code and after (Table 1.1).

Table 1.1. Hawaii's water laws: An evolutionary perspective

Period	Prevailing water rights
Pre-Contact (Early Hawaiian)	"Konohiki" rights
Pre-McBryde (1848-1959)	Appurtenant rights Prescriptive rights "Konohiki" rights Riparian rights Correlative rights
Post-McBryde (1959-1987)	Appurtenant rights Prescriptive rights "Konohiki" rights Riparian rights Correlative rights
Water Code and after (1987 onwards)	Appurtenant rights "Konohiki" rights Riparian rights Correlative rights

The rationale for the classification in Table 1.1 has to do with the landmark case of McBryde v. Robinson, a judicial decision that has significantly affected the evolution and operation of the water rights in Hawaii. The case started in 1959 as the continuation of a dispute between two water users about the diversion of surface waters from the Hanapepe River on Kauai. The crux of the dispute pertained to the private right to control vs. the state's right to control surface water. This long-drawn-out litigation has a chequered history of decisions reversed and appealed a number of times with the final decision rendered by the Ninth Circuit Court of Appeals in 1989, after the state Water Code was enacted in 1987. The Court ruled that "the power to regulate water use and sort out disputes over control of water rights properly belongs with local officials and state courts". This decision was hailed by the then Hawaii State Attorney General as a "landmark victory for Hawaii's people against some of the largest corporations in the state".

Another noteworthy case that has had a major impact on the shaping of water rights in post-McBryde and Water Code periods is Reppun v. Board of Water Supply (1983). In this case, the State Supreme Court directed the Honolulu Board of Water Supply to reduce its groundwater pumping, since the pumping decreased the flow of Waihe' stream, adversely impacting downstream crops.

A brief discussion of each water right noted in Table 1.1 follows. *Konohiki* right, an ancient Hawaiian water right, was formalized during the period of land reform in the mid-1800s. This right designates surface water rights that accompany royal grants of large land areas to the chiefs.

Appurtenant water rights originate from the ancient Hawaiian agricultural practices. These rights confer an entitlement to the quantity of water in use immediately prior to the time the land passed into private ownership. The water rights, basically, are rights to water needed to grow taro on land historically used for this purpose. This customary right became a legal right upon the award of land titles by the Land Commission in 1845.

A *prescriptive* water right is one that is acquired by adverse use. The requirements for acquiring such rights are "that the use of water be actual, open, notorious, hostile to the rightful holder of the water right, and continuous for a statutory length of time, for example, 20 years. If all of these elements are proved, the person has acquired a prescriptive right to use the water"(Castle and Murakami 1991). Such rights are also attached to land, even more vaguely than appurtenant rights.

Riparian water rights are attached only to land through or along which surface water flows. The doctrine is derived from the common law of England. "The owner of the riparian right has only a right of use or a 'usufruct' while the water flows past the owner's property. Each riparian owner is entitled to the natural flow of the stream, diminished only by the reasonable use of others" (Castle and Murakami 1991).

The *correlative* rights pertaining to groundwater use, under common law, implies that overlying lands can legally withdraw water that lies underneath, so long as similar use by adjacent lands over the same aquifer is not injured.

During the pre-contact period (early Hawaiian), the responsibility for the allocation, transfer, and management of water for taro cultivation and other purposes was entrusted by the king to the konohikis or land agents. The konohikis, in effect, served as absentee-landlords and water masters. During this period, non-Hawaiians were not allowed to own water and land. The Great Mahele of 1848 allowed non-Hawaiians, for the first time, to own water and land. During the pre-

McBryde period (1848-1959), a constellation of water rights prevailed: appurtenant rights, prescriptive rights, Konohiki rights, riparian rights (for surface water), and correlative rights (for groundwater). During the period that followed (1959-1987), the same rights that existed during the preceding period, more or less, continued in place.

In reviewing the evolution of water rights in Hawaii, the essential point that needs to be highlighted is the fact that under the ancient Hawaiian system, the concept of private ownership of water simply did not exist. Water was treated as a natural bounty and a public good and the only right that could be ascribed to it was the right to use it, rather than its outright ownership. With the advent of the western settlers in Hawaii, who acquired the right to own land under the Great Mahele in 1848, there arose a major effort to establish private water rights. This effort was aided and abetted by the Konohikis, the Territorial government, private landowners, and assorted other vested interests. Especially noteworthy, in this context, is the long-drawn-out legal battles instigated mostly by the sugar plantations and the big corporations to acquire private rights in Hawaii's public waters. They succeeded in this effort as a result of numerous favourable judicial rulings, until the landmark McBryde ruling in 1989, which, once and for all, settled the question by upholding the state's right in water allocation in Hawaii over private rights.

The 1987 Water Code (Hawaii Revised Statutes, Ch. 174C) represents the culmination of many years' efforts on the part of Hawaii's legislators to enact rules and regulations that would govern the ownership, allocation, and use of Hawaii' surface- and groundwaters. Attempts at any form of regulation, from the outset, were met with strong opposition from the big corporations and oligopolistic landowners. From this perspective, the passage of the Water Code in 1987 has to be viewed as a real breakthrough.

The basic provisions of the Water Code are the following. It provides an administrative mechanism for water resources, consisting of a six-person Commission on Water Resources Management (CWRM), under the jurisdiction of the Department of Land and Natural Resources (DLNR). It designates water management areas where there are indications of potential shortages or decreased quality. All users must file a declaration of use. The Code also provides for regulation of existing wells or diversions and requires permits for the installation of new wells or pumps. In addition, the Code grants the Commission the power to establish rules and gives it jurisdiction of water issues. It allows the state to acquire land for flood control, water management or water-related conservation including streams, beaches, channels, or other measures. Under the Code, the Commission is responsible for gathering hydrologic data and establishing sustainable yields. It allows the Commission to declare water shortages and gives the responsibility to protect water quality and in-stream water uses. Finally, the Code entrusts the Commission with the protection of Native Hawaiian rights.

As it stands, the Code grants considerable power to the state government to regulate water allocation and management through a cumbersome permit and review process. It is unlikely that the landowners are particularly happy with this aspect. However, the Code does allow unlimited duration of use permits; this aspect could soothe landowners fearful of suddenly having their water simply cut off.

Hawaii's Water Code is deficient in many respects. One observer described it as "too little, too late". The major problems with the current Water Code are:

(1) it contains no definition of water rights and water ownership;
(2) provisions governing the sale of water and water transfer (to new uses) are highly restrictive;
(3) there is an overemphasis on 'original use', which curtails flexibility;
(4) the designation of so-called 'Water Management Areas' is fraught with problems;
(5) the Code provides no basis for cost-benefit assessment in monetary terms;
(6) the status of groundwater rights is left unclear.

There is, thus, considerable room for further modification and clarification of the prevailing water rights system in Hawaii.

1.4.2 Political Processes

In this section, a brief historical perspective of the political factors that have governed and shaped freshwater use in Hawaii is given. The politics of water in Hawaii has its roots in a pattern of land ownership unique to Hawaii: oligopoly. Approximately 25 percent of the land in Hawaii is owned by seven large corporations: Alexander & Baldwin, Bishop Estate, C Brewer & Co., Campbell Estate, Dole Food Co., Damon Estate and the Parker Ranch. These big landowners have a pronounced vested interest in the control of Hawaii's waters, since they own most of the sugar plantations and virtually all the pineapple plantations in the state. Sugar and pineapple are both heavily irrigated crops and together, historically (until their recent decline in production), have used about 24 percent of the fresh water consumed.

A review of the pattern of ownership of Hawaii's sugar plantations shows that four major corporations have, historically, owned and cultivated almost 80 percent of Hawaii's sugar plantations. In 1993, 87,000 acres of Hawaii's 126,000 acres of sugar-cane land (69 percent) were irrigated (Hawaiian Sugar Planters Association 1994). The sugar industry applied about 278 million gallons per day (mgd) to cane fields. This amounted to roughly 19 percent of the total water use in the state of Hawaii.

The corporations which dominated Hawaii's sugar industry have over the years sunk their roots far and wide in the ownership, control and appropriation of the state's surface- and groundwaters. The extent of such corporate 'water lordship' is readily apparent from the following facts: "The industry's irrigation systems included about 115 fresh and brackish wells as well as 11 hydroelectric installations, 350 miles of major ditches, and 120 miles of tunnels" (Hawaiian Sugar Planters Association 1994). The replacement cost of the sugar industry's water system in current dollars is estimated to be $1.25 billion (Ibid).

A review of the pattern of fresh water use in Hawaii has shown that agriculture has been the principal user of fresh water in the state, significantly surpassing other uses (64 percent in 1985, 55 percent in 1990, and 54 percent in 2003). This situation has a special bearing, given the fact that sugar and pineapple, largely privately owned and heavily irrigated, have accounted for a substantial part (43 percent) of agricultural water use in the state. This clearly points to the dominance and control of a vital public resource by a few big corporations: a classic case of oligopoly in action.

A closer scrutiny of the make-up of agricultural water use discloses further disturbing patterns. A case in point relates to water consumption by Hawaii's golf courses, many of them under corporate ownership. Water use by golf courses is included in agricultural water use. In 1994, there were approximately 60 privately owned golf courses, which consumed 40 mgd. In 2002, water consumption by private golf courses significantly increased. There were a total of 36 golf courses on the island of Oahu alone. The golf courses consumed, in 2002, an average of 67 mgd (Gopalakrishnan and Cox 2003). Water consumption by golf courses is expected to further increase in the years ahead, based on a projected increase in the number of golf courses. The rapid escalation in the demand for water to meet the growing needs of golf courses has cut into the water available for other uses and has been a continuing source of friction and concern in many communities in Hawaii.

How does the oligopolistic ownership of land and the consequent control of water impact on the state of Hawaii in terms of lost or foregone revenues? What is the monetary impact of the private control of a public resource (in this case, Hawaii's limited fresh water supply with many competing demands, and therefore, clearly constituting an economic good)? To answer this question, Gopalakrishnan et al. (1996) have relied on the concept of a hypothetical water market, and using actual water prices obtained from authentic sources we have developed preliminary estimates of the potential loss of revenues resulting from the oligopolistic control of the state's water resources. This initial effort is not necessarily a precise quantification of the monetary damages to the state, but a reasonable approximation of such losses and could be the basis for further scrutiny, refinement and analysis.

The estimated quantity and value of water used by the sugar plantations of Hawaii during the period 1930-92 are presented in Table 1.2. Using the prices charged by the Honolulu Board of Water Supply ($0.75 per thousand gallons), it was estimated that the value of free water consumed during the 60-year period under review amounts to $6.4 billion involving the use of 8,583 billion gallons of water for the irrigation of almost 8 million acres.

Table 1.2. Estimated quantity and value of free water consumed, based on Honolulu Board of Water Supply (BWS) charges, 1930-1992

Years	Irrigated acres	Water use (billion gallons)	Value of water use ($ million)
1930-39	1,430,930	1,576.000	1,182.00
1940-39	1,337,570	1,473.200	1,104.90
1950-59	1,235,070	1,360.300	1,020.23
1960-69	1,277,820	1,407.400	1,055.55
1970-79	1,207,308	1,312.469	984.35
1980-89	1,119,704	1,164.284	873.21
1990-92	281,632	289.871	217.40
Total	7,890,034	8,583.520	6,437.64

Source: Gopalakrishnan et al. 1996

It should be clear from the above analysis that sugar companies have had the 'free' use of a public resource, namely, the fresh water supply of the state of Hawaii, for several decades largely because of the unclear status of water rights stemming from the absence of a State Water Code clearly delineating the ownership and allocation of the state's surface- and groundwaters. The delay in the enactment of such water legislation to a large measure can be attributed to the political clout of the oligopolists who were in no hurry to write themselves out of the 'free use' they have been enjoying for decades of a public resource.

Our analysis of the close interrelationship between land and water clearly suggests that land ownership without water rights or ready access to water would be of little or no consequence. Given this premise, it becomes readily apparent as to why the big corporations have always wanted a substantial say in the disposition of Hawaii's waters. They have managed to do this for many decades, from the territorial days through to Hawaii's statehood in 1959, dominated largely by the Republicans, and subsequently under largely Democratic administration up until the present.

The big corporations have managed to exercise a substantial measure of influence on the executive and legislative branches of the state government and, in some instances, have reached out even into the judicial arena in eliciting decisions favourable to them with regard to the ownership, allocation and control of water. Key political and administrative positions were routinely held by individuals with direct links to the major corporations (see Gopalakrishnan et al. 1996, for details).

1.4.3 Water Administration

The single most important agency responsible for the allocation and management of water in Hawaii is the state Department of Land and Natural Resources (DLNR). More specifically, the responsibility is entrusted with the Commission on Water Resources Management (CWRM). The broad mission of the Water Commission is to protect and enhance Hawaii's water resources through judicious and responsible management. The Commission consists of six members (Commissioners). Two are ex-officio members and the other four are political appointees of the governor. Although autonomous, it is administratively housed in the Water Resource Management Division of DLNR.

Another important state agency is the state Department of Health, especially its Safe Drinking Water Branch, which is responsible for monitoring water quality. The state Department of Agriculture, especially its Pesticides Branch, is also concerned with water quality.

Federal agencies involved in the management and conservation of water in Hawaii include the Environmental Protection Agency (EPA), U.S. Geological Survey (USGS), and the Natural Resources Conservation Service (NRCS). In this connection, the largely federally funded University of Hawaii Water Resources Research Centre (WRRC), whose main mission is water-related research and its dissemination, also deserves mention.

The major local agency supplying water to the growing population of Honolulu, the largest city and the major population centre of the state of Hawaii, is the Honolulu Board of Water Supply. Especially important is its role in the supply of municipal, residential, and visitor industry water needs.

At the local level, there are also a number of programmes concerned with water quality and conservation. These include work done by the county extension agents, irrigation districts, wellhead protection programmes, associations of agricultural producers, such as Hawaiian Sugar Planters Association (HSPA) and Hawaiian Pineapple Growers Association (HPGA), and owners and operators of private ditches.

Agencies responsible for water allocation and management in Hawaii suffer from serious constraints, most notably at the state level. The Water Commission currently lacks autonomy. To ensure autonomy, the Commission, currently housed in DLNR, should be moved out of it and set up as a separate agency, entrusted with the responsibility for all aspects of water administration. The Commission, presently, is severely understaffed and under-funded, especially given its statewide responsibilities for water allocation and management. Thus, a clear need exists for legislative appropriations to significantly augment the Commission's manpower and financial resources.

Taking into account the changing configuration of water demands and the potential that it generates for escalating conflicts among different water uses and users, the necessity to elicit and incorporate the concerns and suggestions of multiple stakeholders becomes compelling. In Hawaii today, there is no specific mechanism to achieve this goal, other than the routine public hearing process on contested water issues, which is far from adequate. We can thus conclude that a major revamping of Hawaii's water administration agencies is essential to enhance the effectiveness of water allocation, planning, and management in the state.

1.5 Performance of Hawaii's Water Institutions: A Case Study

The purpose of this section is to review and assess the effectiveness and efficiency of the water institutions that are currently in place in Hawaii in the planning, allocation, and management of water resources. This is accomplished by a detailed case study of a heavily contested case involving water allocation by the state among multiple stakeholders. All three components of water institutions viz. water laws, political processes, and water agencies, noted in the previous section, figure prominently in this case study and thus it provides an excellent opportunity to examine and evaluate the performance of Hawaii's water institutions in terms of well-defined criteria.

This case study deals with the 1997 allocation of the Waiahole Ditch waters on Oahu, the major economic and population centre and visitor destination of Hawaii. The Waiahole ditch and tunnel system was built in 1916 by the Oahu Sugar Company for transporting water from the wet, windward side to the arid leeward plain of Oahu for irrigating its extensive sugar plantation spread over some 4,000 acres. For almost 80 years, approximately 27 mgd of water was being diverted through this 25-mile tunnel for the exclusive use of Oahu Sugar Company. The closing down of sugar operation at the end of 1995 freed this water for alternate uses.

1.5.1 The Problem

The central issue in the Waiahole ditch controversy boils down to the question of how best to reallocate this water among competing and often conflicting demands. The Hawaii Commission on Water Resources Management (hereinafter called the Water Commission) became responsible for the reallocation of the surplus 27 mgd of water among uses and users claiming a share in it. There are many stakeholders in this water bonanza. These include, on the leeward side, large landowners (Bishop, Castle, Campbell and Robinson Estates), and big resort-, golf-course-, and housing-developers. On the windward side, the claimants include small farmers, community associations, neighbourhood boards, Native Hawaiian Associations, and assorted environmental groups.

The Water Commission, as the agency entrusted with the management of the state's water resources, assumed the responsibility for reallocating the surplus Waiahole water. The two major competing water users, as noted above, were the urban developers of leeward and central Oahu, on the one hand, and a coalition of farmers and environmentalists of windward Oahu, on the other. The urban developers have a close affiliation with Hawaii's big corporations and they stand to gain enormously if substantial quantities of water are made available to them at throw-away prices.

The Water Commission, after more than three years of deliberations, announced in December 1997 its decision with respect to the reallocation of 27 mgd of Waiahole water among the contending parties as follows: 14.03 mgd for the leeward side and 12.97 mgd for the windward side. The 14.03 mgd of water allocated to the leeward side included 12.22 mgd for agricultural uses and 1.29 mgd for former Oahu Sugar Company sugarcane lands currently used for diversified agriculture (nearly 4,000 acres at 2,500 gd/acre), and 2.2 mgd for Castle Corporation's agricultural lands. The windward allocation by the Water Commission included a release into the windward streams a total of 12.97 mgd, out of which 6.77 mgd remained water available for offstream leeward uses. The amount of water released to windward streams was significantly lower than was requested by the windward parties.

1.5.2 Analysis and Critique

There is no mention in the Water Commission's report of the specific criteria it used to reallocate water among the different uses and users. Absent from the Commission's report are explicit and clear discussions of: (1) Water rights issues; (2) Impact on sustainable water use; (3) Impact on water quality; (4) Impact on instream uses; (5) Economic impacts; (6) Impact on water conservation; and (7) Impact on watersheds. The exclusion by the Commission of these criteria for water allocation poses troubling questions as to whether the projected increases in water demand for the different sectors of Hawaii's economy have been considered in the allocation of water.

1.5.3 Water Rights Issues

A group of water users challenged in the State Supreme Court the Commission's allocation of water among different users, shortly after the final allocation was announced in December 1997. Of particular concern to the plaintiffs was the large quantity of water allocated to agriculture on the leeward side. It was argued that the amount allotted for agricultural use by the leeward landowners was excessive and this could well have been the result of the Water Commission using an arbitrary water per acre need for diversified agriculture as the allocational basis.

A second major concern had to do with the Water Commission's rationale and approach toward the allocation of water for in-stream uses. The State Supreme Court took strong exception to the Commission's allocation. The Court ruled that "without any proper findings as to the actual requirements for in-stream purposes, or the reasonableness of offstream diversions relative to these requirements, the Commission effectively assigned to windward streams the water remaining after it had approved the bulk of the offstream use permits" (Waiahole Water Case 2000).

The State Supreme Court concluded that the Water Commission had overly narrowly applied the Public Trust Doctrine, the basic legal principle governing the allocation and appropriation of natural resources, including water, in its deliberations and decisions, to the detriment of many parties. The Court urged the Commission to take into account issues of intergenerational equity, ecological and environmental impacts, Native Hawaiian Water Rights, and other related concerns in its efforts to reapportion the Waiahole water. The State Supreme Court therefore revoked several of the Water Commission's decisions and remanded it to revisit and revise its earlier decisions consistent with the Court's decision (Waiahole Water Case 2000).

1.5.4 Impact on Sustainable Water Use

It is estimated that the demand for water in Hawaii will increase by about 206 mgd in 2010 from the current level of consumption. The demand for domestic water use alone will be 116 mgd in 2010, representing an increase of 30 mgd (35 percent) over the water consumption in 1990 (Malla and Gopalakrishnan 1997). The commercial-industrial sector of Hawaii is another important consumer of the state's water supply, with a clear potential for significant additional water demand in the years to come. This is evidenced by the fact that commercial and industrial water use increased by a dramatic 60 percent during 1985-1990 in the state of Hawaii (Malla and Gopalakrishnan 1999). The visitor industry's water consumption is also expected to go up markedly (Gopalakrishnan and Cox 2003).

Oahu's sustainable yield estimate for groundwater was revised to 465 mgd from 495 mgd by the Water Commission in 1991. The demand projections estimate that Oahu will be approaching the sustainable yield limits for the island within the next 20 years. Nevertheless, the water withdrawal in two of the main aquifer systems is already 85 percent or more of the sustainable yield (85 percent in Pearl Harbor and 88 percent in Honolulu) (Malla and Gopalakrishnan 1995). Thus, it becomes imperative to undertake a careful review and analysis of the potential impacts of any proposed water allocation on sustainable water use in the state of Hawaii. Regretfully, no attempt was made by the Commission to examine this aspect.

1.5.5 Impact on Water Quality

Approximately 85 percent of the drinking water in Hawaii comes from public supplies with a groundwater source (Hagemann and Fukunaga 1995). No other state in the nation is so heavily dependent on groundwater for its rapidly growing water needs (Malla and Gopalakrishnan 1997). Given this extraordinary situation, the case for preserving the quality of Hawaii's groundwater supplies becomes compelling.

There have been many instances of water contamination in Hawaii from non-point sources, virtually all of them from agricultural chemicals (Lau 1987). Given the preponderance of water allocation for agricultural use, the Water Commission should have used due diligence in assessing probable water quality impacts from the use of agricultural chemicals in diversified agriculture. The potential for surface water pollution from residential, industrial, and urban developments triggered by the reallocation of water should also have been carefully examined by the Water Commission. This is all-the-more important in the case of Hawaii's fragile tropical environment which is so vital for sustaining the visitor industry, especially its growing segment of eco-tourism. Unfortunately, the Water Commission made no attempt to examine and assess the potential water quality impacts.

1.5.6 Impact on In-Stream Uses

The windward farmers contend that a return of water is crucial for the restoration of shrunken streams and the replenishment of stream-fed estuaries where fish breed. It has been maintained that the stream ecosystem and the traditional taro and rice farming in the Waiahole-Waikane areas dried up 80 years ago when the ditch was constructed to divert water to central Oahu.

Monitoring of increased stream flow has shown that a category of freshwater fish, O'opu, has made a comeback and benefited the Kaneohe Bay where the stream empties (Matsunaga 1995). Kaneohe Bay once had 70 mgd of freshwater flowing into it by way of stream, 50 percent of which no longer enters the bay. Therefore, it stands to reason that the availability of water in windward streams, among other things, will improve the habitat for near-shore fisheries, provide opportunities for aquaculture, enhance stream habitat, and increase valuable wetland habitat for the endangered birds (Reppun 1994a).

1.5.7 Potential Economic Impacts

Two possible scenarios of water allocation have been put forward, one proposing retaining water for the leeward side, while the other suggesting returning the water to the windward streams.

Leeward Oahu

The interests on the leeward side include Amfac (former owner of the Waiahole Water Company), large landowners—the Bishop, Campbell and Robinson Estates—and their neighbours who are supported by the resort, residential, and golf course developers, as well as the state Departments of Land and Natural Resources and Agriculture.

It has been argued that water should be retained on the leeward side for diversified agriculture and urban development. The windward side receives heavy rainfall compared with dry central Oahu. Land requesting water in central Oahu totals 9,000 acres (Paty 1994). Over 4,000 acres of land can be put into diversified agriculture consistent with the current state policy. Crops such as tomato, bell pepper, potato, herbs, wonbok, and more can be grown to meet local demands (substituting imports) and also for export. It is estimated that diversified agriculture can generate one job per 20 acres against one for 30 acres in sugarcane (Paty 1994), helping to absorb jobs displaced by sugar workers. Thus, there is a potential to create about 450 jobs on the leeward side.

Water can also be used in urban areas for landscaping, industrial operations, golf course irrigation, and fire protection. Some advocating more water for the leeward side argue that any water allocation decision should be based on General Plans for the island of Oahu. Leeward Oahu is envisioned to grow significantly in the Oahu General Plan, while the plans for windward Oahu call for virtually no growth and emphasize its rural character (Davidson and Callies 1995).

Windward Oahu

Those supporting water restoration to the windward side include the Waiahole-Waikane Community Association, the Kahuluu Neighborhood Board and Ka Lahui Hawaii, a Native Hawaiian group (Matsunaga 1995). However, it was suggested by the Water Commission that the freed water maybe enough to meet the demands of both sides of the Koolau mountains (Wagner 1994).

On restoration of water to the windward side, it has been claimed that about 180 acres of land can be re-established under rice crop and 400 acres under taro cultivation. Currently, taro is being cultivated on 10-15 acres (Barayuga 1994). It has also been suggested that 1,500 additional acres can be made productive provided the water is brought back (Reppun 1994 b).

The windward farmers argue that thousands of acres of central Oahu land are vacant and therefore do not need water. Besides, there are possibilities of developing alternate sources in leeward/central Oahu, including millions of gallons of treated sewage effluent for irrigation. Since a substantial amount of taxpayer money is involved in treating the effluent, it seems only logical to make use of this water for irrigation. Another alternative is the underground pumping of water.

The windward farmers point out that the increased flow of water has facilitated the cultivation of taro crop. There are now more commercial baitfish and shrimp. The future water requirements for agricultural and aquacultural uses in windward Oahu are projected to reach 34.70 mgd in the years ahead.

1.5.8 Water Conservation

Given the escalating demand for and shrinking supplies of both surface and groundwater supplies in Hawaii, the potential for conservation should be a key factor in any proposal involving water reallocation. It appears that the Oahu Sugar Company has not been able to fully utilize for many years the 27 mgd it had contracted for. Instead, it was routinely dumping as much as 18 mgd of water in dry gulches, pastures, and ditches, a practice seriously compromising efficient water use and water conservation (Vickers 1994). Also, the potential for using treated effluent for the purpose of recharging the Pearl Harbor aquifer should be carefully examined (Environment Hawaii 1994). It is also estimated that 25 mgd of treated sewage effluent can be made available for possible irrigation purposes in leeward Oahu.

1.5.9 Impact on Watersheds

Any scheme involving a major reallocation of water should carefully consider and document the impact of the proposed diversion on the watershed as a whole, as opposed to its impact on the immediate vicinity. The emphasis today in the context of water allocation and transfer is on the macro impacts on the area served by the whole watershed instead of the micro impacts on specific areas immediately affected, as was the case in the past. The hydrology of the watershed should be at least as important as the economic impact in charting the course of water reallocation. This is especially important since sustainable water use would dictate a rate of depletion no higher than the rate of replenishment. Also, an examination of the impacts on ecology, the environment, and aesthetics of the watershed, along with the consideration of irreversibility, should be part of the calculus of decision-making.

1.5.10 Conclusions

This case study has shown how imperative it is to have explicit and well-defined criteria in place in order to equitably allocate a public good, such as water resources, among myriad competing demands. Adherence to these criteria should ensure access and availability of water to the different sectors, including the disenfranchised minority groups (in this case the Native Hawaiians), thus meeting the fundamental tenet of allocational justice of a public good.

Arbitrary allocation of water by bureaucratic agencies has been widely recognized as the principal cause of conflicts among stakeholders. In most instances, the absence of a framework or system based on well-defined criteria, and the consequent lack of accountability, has been the main source of such arbitrary action. The upshot has been the proliferation of conflicts among uses and users involving long-drawn-out negotiations and other conflict resolution mechanisms with prohibitive transactions costs. The application of the proposed seven-points framework has the potential for rational water allocation and a substantial reduction in costly water use conflicts and litigation.

1.6 Hawaii's Water Institutions: A Case of Institutional Entropy

A review of the performance of Hawaii's water institutions during their evolutionary phase and in recent times, points to the pervasive presence and operation of entropy and its debilitating effects. Institutional entropy has resulted in a significant reduction in the ability of institutions to meet and address the changing and challenging needs, or the changing configuration of water uses and users. In essence, this has led to a diminished capacity on the part of institutions to effectively perform their responsibilities, as the discussion below shows.

First, the performance was affected by lack of flexibility. The many judicial rulings handed down by the courts with respect to the allocation of water, especially during the pre-McBryde years, have routinely favoured private rights in water, overlooking the idea of water as a public good embodied in the ancient Hawaiian system. More recently, as the Waiahole Ditch case study illustrates, the lack of flexibility on the part of the Water Commission in the interpretation and application of an appropriate water law resulted in a flawed allocation of water among competing uses and users for the Waiahole surplus water, a decision vacated by the State Supreme Court in its landmark ruling in 2002. The Commission's allocation of the Waiahole water among contending parties also showed a lack of flexibility in making allowance for the newly-emerging water uses (e.g. environmental) in Hawaii.

Second, the principle of autonomy has been compromised in a number of decisions pertaining to the ownership, allocation, and management of water in Hawaii. For instance, CWRM, the principal agency responsible for water administration is housed in the Department of Land and Natural Resources (DLNR), and has the director of DLNR as its chair and, therefore, it clearly lacks autonomy. In fact, the Commission's allocation of Waiahole water among multiple stakeholders has been viewed by many water experts and analysts as favouring certain parties close to the executive branch. Given the politics of water in Hawaii discussed in 4.2 above, it should be clear that the climate was not conducive for independent decision-making. The long and tortuous history of the difficulties encountered in securing legislative approval to a State Water Code as late as 1987, attests to the corporate greed and power that stymied the concept of water as a public good. This alliance of convenience by the different interested parties accelerated the growth and spread of entropy in the institutions entrusted with the task of optimal water governance.

Third, the decisions taken by the water agencies in the past, as well as in more recent times, suggest either a lack of access to or disregard for the full range of pertinent information and data needed for a careful and comprehensive analysis. For example, although the damage to in-stream uses in windward Oahu due to the age-old diversion of Waiahole water has been well-documented, there is no evidence of the Commission factoring this into its deliberations on water allocation for in-stream uses. Another example is the arbitrary allocation of excessive, per-acre, irrigation water for diversified agriculture, unsupported by scientific data. These clearly illustrate the prevalence of institutional entropy detracting from informed decision-making.

Fourth, lack of cultural calibration, noted in 1.3 as a key deterrent to optimal decision-making, is evidenced by the scant attention paid by the Commission to the legitimate demands of windward taro and rice farmers and in-stream users, further contributing to institutional entropy. Also, not considering adequately issues of water quality, water conservation, and water sustainability, all of which are emerging as key areas of concern to the people of Hawaii, in the allocation of water, further attests to a departure from cultural calibration, a key element of institutional durability.

This review of the performance of Hawaii's water institutions, strongly points to the presence of many major constraints that have proved to be a fertile ground for the rapid growth and spread of lethal doses of system-choking entropy. To rid the system of such corrosive growth, quick, decisive, and effective action is called for. There is growing evidence of popular disenchantment with Hawaii's water institutions, which for all intents and purposes, have remained the product of a monumentally moribund system representing a long-vanished era. Thus, it appears that the time is opportune to identify, introduce, and implement institutional innovations in the planning, allocation, and management of Hawaii's water resources.

1.7 Future Directions

In this section, a few suggestions for revamping Hawaii's water institutions are presented. These are based on the findings and conclusions that have emerged from the foregoing analysis. The overarching goal of these recommendations is to significantly reduce the entropy-producing factors and features of Hawaii's water institutions and thus strengthen and augment their ability to perform effectively.

Competition for water, as our discussion has shown, is looming large as a major problem in the context of Hawaii's water planning and management. It is therefore vital to develop mechanisms to accommodate, efficiently and equitably, the changing configuration of water uses in the state. The most pressing need in this connection is to examine the feasibility of developing water markets and water pricing so that water users can be charged fair market prices.

Another important change has to do with the interpretation and application of Hawaii's water laws appropriately by the Water Commission. In this connection, the Commission has already received guidance from the State Supreme Court in its ruling handed down in 2002, with respect to the allocation of Waiahole waters. The justices ruled in this case that the pre-eminent water doctrine that governs water ownership and allocation in Hawaii is the Public Trust Doctrine (for a detailed discussion of the Public Trust Doctrine, see Sax 2001; Powers 1995; Copeland 1991; Rieser 1991), and not variations of the common law Riparian Doctrine or the Doctrine of Prior Appropriation. The Commission's future actions with respect to water allocation should be guided by this landmark ruling.

The Water Commission, in making decisions on the ownership, allocation, and management of Hawaii's water resources, should use well-defined criteria of efficiency, equity, and sustainability. Issues of water quality and water conservation should be factored in along with considerations of economic development. Also, a macro approach involving the watershed as a whole should be the basis for decision-making.

Given the centrality of water, it is important to have a single, autonomous agency responsible for most, if not all, aspects of Hawaii's water administration. For instance, decisions pertaining to water allocation, conservation, and quality can be best made by a single fully-staffed and well-funded state agency with a clear understanding of the ramifications of the decisions made. Also, a single agency can avoid jurisdictional overlapping and the consequent turf wars and delays in timely decision-making.

The current Water Code of Hawaii is in need of serious revamping. Corporate politics appears to have influenced several provisions of the Water Code. This situation has to be altered to ensure the orderly development and wise harnessing of the state's dwindling water supply to meet escalating demands. An important area of reform, in this context, has to do with key appointments to relevant state agencies and commissions responsible for making major decisions on the ownership, control, allocation, and appropriation of water resources. Individuals with vested interests, corporate or otherwise, should not be allowed to serve on such decision-making bodies.

Another important issue that merits careful attention pertains to Native Hawaiian water rights. The 1987 Water Code has dodged this issue. There have been no serious attempts on the part of the Water Commission, the State Legislature, or the executive branch to actively address this issue as expeditiously as possible. Mobilization of a major effort to address this issue is long overdue.

In summary, the key issues to be addressed in charting the future course of Hawaii's water development include: application of the Public Trust Doctrine as the basis for water ownership, management, and allocation decisions; development of a single autonomous agency responsible for water administration; provision for the development of water markets and pricing to ensure efficient and equitable water use; water use allocation and management using clearly defined criteria; and the recognition of Native Hawaiian water rights. Deliberations on these and other issues should proceed in a setting free from the politics of special interest groups.

1.8 References

Aoki M (2001) Toward a comparative institutional analysis. Massachusetts Institute of Technology, Cambridge, MA

Barayuga D (1994) Developers, golf courses may get windward water; a company's proposal is opposed by Waiahole farmers. Honolulu Star Bulletin, 20 June

Beaumont P (2004) Water institutions in the Middle East. In: Gopalakrishnan C, Tortajada C, Biswas AK (eds) Water Institutions: Policies, Performance and Prospects. Springer, Berlin

Castle J, Murakami A (1991) Water rights. In: Mackenzie, K.M (ed) Native Hawaiian rights handbook. Native Hawaiian Legal Corporation, Honolulu, pp 153-159

Commission on Water Resources Management (1990) Oahu water management plan. State of Hawaii Department of Land and Natural Resources, Honolulu

Commons JR (1995) Legal foundations of capitalism (Rev Ed). Transaction Publishers, New Brunswick, NJ

Copeland S (2001) Proceedings of the 2001 symposium on managing Hawaii's Public Trust Doctrine, Honolulu

Davidson D, Callies DL (1995) Base water decision on existing plans. Honolulu Advertiser, 27 July

Derrickson SAK, Robotham MP, Olive SG, Evensen CI (2002) Watershed management and policy in Hawaii: coming full circle. Journal of the American Water Resources Association 38:563-576

Gardner DG (2003) Weakening water rights and efficient transfers. International Journal of Water Resources Development 19:7-19

Guiasu RC, Guiasu S (2003) Entropy in ecology and ethology. Nova Science Publishers Inc, Hauppauge, NY

Gopalakrishnan C, Cox LG (2003) Visitor industry water demand: the case of Hawaii. International Journal of Water Resources Development 19:29-35

Gopalakrishnan C, Malla P, Khaleghi GH (1996) The politics of water in Hawaii: an institutional appraisal. International Journal of Water Resources Development 12: 297-310

Greven A, Keller G, Warnacke G (eds) (2003) Entropy. Princeton University Press, Princeton, NJ

Gunatilake HM, Gopalakrishnan C (2002) Proposed water policy for Sri Lanka: the policy versus the policy process. International Journal of Water Resources Development 18: 547-564

Hagemann M, Fukunaga G (1995) A demonstration project to estimate the vulnerability of central Oahu's groundwater to anthropogenic contaminants. Environmental Protection Agency, Washington D.C. and State of Hawaii Department of Health, Honolulu

Honolulu Advertiser (1995) Editorial. Waiahole ditch: job for planners, not lawyers, 5 July

Howe CW (2004) Property rights, water rights, and the changing scene in western water. In: Gopalakrishnan C, Tortajada C, Biswas AK (eds) Water Institutions: Policies, Performance and Prospects. Springer, Berlin

Huffaker R, Whittlesey N, Hamilton R (2000) The role of prior appropriation in allocation of water resources into the 21st century. International Journal of Water Resources Development 16:265-273

Karmeshu J (2003) Entropy measures, maximum entropy principle, and emerging applications. Springer-Verlag, New York

Kirch PV (1985) Feathered goods and fishhooks: an introduction to Hawaiian archaeology and prehistory. University of Hawaii Press, Honolulu

Lau L S (1987) An overview of Hawaii's water resources with reference to organic chemicals In: Toxic organic chemicals in Hawaii's water resources, HITAHR Research Extension Series 086, University of Hawaii, Honolulu

Malla P, Gopalakrishnan C (1999) The economics of urban water demand: the case of industrial and commercial water use in Hawaii. International Journal of Water Resources Development 15: 367-374

Malla P, Gopalakrishnan C (1997) Residential water demand in a fast-growing metropolis: the case of Honolulu, Hawaii. International Journal of Water Resources Development 13:35-51

Malla P, Gopalakrishnan C (1995) Conservation effects of irrigation water supply pricing: a case study from Oahu, Hawaii. International Journal of Water Resources Development 11:233-242

Matsunaga M (1995) On both sides of Koolaus, a need for Waiahole water. Honolulu Advertiser, 26 June

McBryde Sugar Company v. Robinson 54 Haw. 174, 504 P.2d 1330 (1973)

Nabli MK, Nugent JB (1989) The new institutional economics and development. North Holland Publishers, Amsterdam

Nickum JE (2004) Uphill flow of reform in China´s irrigation districts. In: Gopalakrishnan C, Tortajada C, Biswas AK (eds) Water Institutions: Policies, Performance and Prospects. Springer, Berlin

North DC (1989) Institutions and economic growth: an historical introduction. World Development 17:1321-1322

Ostrom E, Schroeder L, Wynne S (1993) Institutional incentives and sustainable development: infrastructure policies in perspective. Westview Press, Boulder, Co

Paty B (1994) Should windward water still go to central Oahu? Yes. Honolulu Advertiser, 27 November

Pejovich S (1995) Economic analysis of institutions and systems. Kluwer Academic Publishers, Dordrecht

Powers JP (1995) Reinvigorating natural resources damages actions through the Public Trust Doctrine. New York University Environmental Law Journal 4:418-448

Reppun C (1994a) Waiahole water: it's for windward. Honolulu Advertiser, 14 August

Reppun C (1994b) Should windward water still go to central Oahu? No. Honolulu Advertiser, 27 November

Reppun v. Board of Water Supply 65 Haw. 531, 656 P.d 57 (1982), cert. denied, 471 U.S. 014, 105 S. Ct 2016, 85 L Ed 2d 298 (1985)

Rieser A (1991) Ecological preservation as a public property right: an emerging doctrine in search of a theory. Harvard Environmental Law Review 15: 393-434

Rifkin J (1981) Entropy: into the greenhouse world. Bantam Books (Rev Ed), New York

Ruttan VM, Hayami Y (1984) Towards a theory of induced institutional innovation. Journal of Development Studies 20: 203-223

Saleth RM (1996) Water institutions in India: economics, law, and policy. Commonwealth Publishers, New Delhi

Sax J (2001) Proceedings of the 2001 symposium on managing Hawaii's Public Trust Doctrine, Honolulu

Vaidyanathan A (1999) Water resource management: institutions and irrigation development in India. Oxford University Press, New Delhi

Vickers M (1994) Water from Waiahole taken, then dumped. Honolulu Star Bulletin, 20 May

Waiahole Water Case, In the Matter of the Water Use Permit Applications, 94 Haw. 97 (August22, 2000)

Wagner P (1994) On tap; a sticky battle for old Waiahole water, windward wants to plug ditch's flow to leeward lands. Honolulu Star Bulletin, 20 May

Institutions for Resources Management: A Case Study from Sri Lanka

Asit K. Biswas

2.1 Introduction

In the area of water resources management, in depth analyses of national and international institutions have been a most neglected issue. This, in one sense, is not surprising since all the institutions would like to have their successes and strengths to be highlighted, but very few would like to have their weaknesses, constraints and failures to be critically examined in public. Thus, most institutions would not like, and in fact actively discourage, to have their overall performance to be analysed frankly and objectively. Accordingly, they mostly prefer not to cooperate with independent and competent analysts, especially if such assessments are unlikely to remain confidential. Thus, while some pseudo-evaluations are currently available, which mostly show the institutions concerned and their programmes in a positive light, really serious assessments of institutional performance are conspicuous because of their absence.

This is an unwelcome state of affairs, especially as in most countries of the world, it is the institutions that formulate water policies, or at the very least, are primarily responsible for the formulation of such policies through discussions and consultations with the other appropriate stakeholders. It is also the institutions which for the most part decide the processes through which water policies are to be formulated. They often have the authority (de facto, if not de jure) to guide not only the policy formulation processes, but also often decide, directly or indirectly, the direction and the main components of the policies that are to be formulated.

Once the policies are formulated, it is the institutions that are mostly responsible for their implementation. Also, the subsequent functioning of the institutions is dependent upon the policies whose formulation is assisted by the very same institutions. Thus, in the final analysis, it is the institutions that are mostly responsible to ensure the good governance of the water sector. Without efficient and functioning institutions, rational management of water resources is simply not possible. Furthermore, if the water institutions are not efficient, even the best approaches, processes, techniques and technologies that are available at present can at best make marginal contributions to proper water resources planning and management. Good water governance and management is simply not possible without capable and efficient institutions.

It is thus an anomaly that even though the institutions play a central and pivotal role in efficient water management, they are rarely analysed and assessed objectively and regularly. It is essential to know why certain water institutions work properly, but others do not. It is equally necessary to know their individual strengths and weaknesses, how their strengths could be maximized and the weaknesses could be minimized, what constraints they face and how these constraints could be overcome cost-effectively and in a timely manner. Comparative studies of different institutional structures and their performance need to be carried out to find out which institutional framework is likely to be most suitable for a specific location, at a specific time, and why. The institutional knowledge-base at present is very limited, even though the need for good governance for the water sector has been increasingly recognized to be an essential requirement during the post-1995 era. A good and effective institution is undoubtedly at the heart of any rational water resources planning and management process.

2.2 Public and Private Institutions

A major issue of contention during the past decade has been the role of the private sector institutions in water resources management. In reality, however, nearly all the discussions have been focused only on the performance of the private sector institutions associated with water supply and wastewater management. This is inspite of the fact that the private sector institutions have also been involved in the hydropower sector (for example, in countries like Turkey and India) and also in construction and operation of desalination plants in several countries in and around the Middle East. There has been only limited involvement of the private sector institutions in agricultural water management, as in Chile. Overall, discussions of the desirability and the performance of the private institutions in managing water in agricultural, hydropower and desalination sectors have received scant attention thus far.

Even after some 10 years of intensive, and often heated, discussions on the roles of private institutions to manage water supply and wastewater management, there is no general consensus in the water profession at present as to which alternative works best, where, under what conditions and why. Not surprisingly, it was one of the two issues (the other one being large dams) that generated maximum interest and controversy during the Second World Water Forum in The Hague in 2000, and also during the Third World Water Forum in Japan in 2003. The discussions on the overall desirability and performance of private sector institutions were haunted by dogmas, vested interests, pseudo-scientific analyses, personal biases and limited understanding of the water problems and needs of developing societies, and how these can be best addressed in a timely, long-term, cost-effective, and socially-acceptable manner.

The views of individuals and organizations promoting or opposing the involvement of the private institutions in water management are often deeply held, and neither side has bothered much to listen carefully, let alone consider, the views and concerns of the other side. Thus, the discussions of this issue, encompassing both proponents and opponents at various international fora, have mostly ended up as dialogues between the deaf.

There is no question that during the late-1990s, there were unrealistic expectations that the private sector institutions will dramatically improve the management of the public utilities by bringing in new management techniques, well-trained people, new capital and advanced technology. It was expected to solve most, if not all, problems of the public sector institutions. Major multilateral organizations like the World Bank and the Asian Development Bank have in recent years pushed strongly for the involvement of the private sector institutions in water management.

The report of the World Commission on Water for the 21st Century (2000) made several comments on the role of the private sector institutions to manage water in the future. It pointed out that:

- Adequate incentives should be provided to the private sector to contribute where it is well-equipped to do so.
- The private sector will bring neither its money nor its management skills and know-how, unless it can operate in a predictable, transparent regulatory environment, and unless it can get a reasonable return on its investments, without undue political interference.
- Private sector can bring in additional financing to the water sector, which public sector alone cannot provide.
- Private sector can considerably improve the current poor technical and financial performance of most water utilities in the developing world.

The euphoria that the private sector institutions would significantly improve water development and management practices and processes in developing countries is now basically over. Sober realism, and post-2000 developments, indicate that:

- The private sector institutions currently serve about 3-6 percent of urban water consumers in the developing world. This percentage is likely to increase in the coming years, but at a much slower rate than that was anticipated only four years ago.
- Under all foreseeable conditions, by the year 2020, it is highly unlikely that more than 15 percent of the domestic urban consumers in the developing world will receive their water and wastewater services from private sector institutions.

Hence, the real issue at present is not that the private sector institutions will not play important roles in the future (they for sure will in many different forms), but how can the performance of the existing public sector institutions be significantly improved, because even by 2020, they are likely to provide the services to the vast majority of the domestic urban consumers.

There is some evidence at present to indicate that at least some of the international organizations felt that it would be very difficult to reform the public sector water institutions within a reasonable timeframe because of their inherent political, institutional and economic constraints. Cultural conditions of a country affect the performance of its public sector institutions. This is because national culture, to a very significant extent, determines the performance efficiencies of public sector institutions, which are run by people, rather than the prevailing laws and regulations. Furthermore, in all countries, it is almost impossible to conceive that only the water institution could be significantly more efficient that the other public sector institutions on a long-term basis.

Many multilateral and bilateral institutions felt that an easier solution would be to handover the management services to private sector institutions, which could work under new rules and regimes that would overcome many of the current constraints faced by the public sector institutions, which reduce their operating and management efficiencies. It would be more difficult and time-consuming to get the public sector institutions to accept the new rules and working practices which could increase their operating and management efficiencies very significantly. Thus, handing over management to the private sector seemed to be a simpler and more attractive option than reforming the public sector institutions.

A good example is the current situation in Morocco. If any public sector institution wishes to increase water prices even by a very small amount, such requests have to be debated and approved by the National Parliament. Thereafter, the King would have to approve the price increases. The process thus is complex, time-consuming, and the results are not always certain or predictable. For example, some municipalities which have requested approvals for increases in water prices have not received the necessary authorization in three to five years. During this period, they need even another price rise simply to keep up with the inflationary pressures.

In contrast, the private sector in Morocco can increase water prices at present within specific predetermined bands. Thus, the playing field is no longer level between the public and private sector institutions in many different parts of the world.

Based on extensive analyses of the performance of the private sector institutions in Middle East and North Africa (Al Baz and Biswas 2003), and Latin America (Tortajada and Biswas 2004), following conclusions can be drawn:

1. There are many forms of involvement of private sector institutions in the water sector. These could range from outright sales of assets of public water institutions to the private sector, as was the case for England and Wales, to provision of management concessions to run water supply and wastewater collection and treatment facilities over a fixed number of years (current concessions range from two to 60 years), to outsourcing of numerous specific activities which the private sector can do more economically and efficiently compared to the public sector. Since England and Wales sold all their assets outright to the private sector in 1989, no other country has followed this model. More than a decade after this privatization in England and Wales was completed, there is currently no agreement amongst water professionals as to its actual impacts on the consumers and on the quality of services provided. The assessments available at present range from highly favourable to equally highly deplorable. In contrast to outright sale of assets, the use of management contracts with appropriate private sector institutions for a specific period of years has proliferated during recent years. Here again, no universal judgment can be made. Some concessions have been very successful, but equally others have been dismal failures. Results have sometimes varied even within a single country (for example, in Morocco, Casablanca could be considered to be a success but not Rabat), and in one instance at least within the same metropolitan area (half of Manila worked but the other half did not). There could be a time dimension to the effectiveness of the private sector institutions as well. For example, Buenos Aires was a reasonably good example of private sector involvement, until the economic meltdown in Argentina changed all the prevailing boundary conditions. Thus, the current analyses indicate that there is no single model of institutional participation that would be most optimal to all cases within a single country, let alone for the entire world.
2. After rapid expansion in the award of concessions to manage water supply and wastewater systems to private institutions in recent years, the rate of awards of new similar concessions has slowed down perceptibly, from about the year 2001. Near-term prospects for acceleration of the rate of awards of new concessions do not look encouraging.
3. Competitive pressure from multinational water institutions has improved the performance of public sector institutions in many developed as well as developing countries. This is an important benefit that the water professionals have mostly ignored. It is likely that the performance of many of the public sector institutions are likely to improve significantly in the coming years, probably at a much higher rate than that was the norm during the past 20 years. Without the threat of private sector competition, it was highly unlikely that the performance of public sector institutions would have improved in such a remarkable fashion in recent years, and over such a short period of time.
4. Performance of public and private sector institutions should not be generalized. By most criteria, the best water utility in the world continues to be Singapore, a public sector institution, but with considerable autonomy. Even within private sector institutions, performance has varied from one city to another in one country, and also could vary over time even in the same city. Thus, any objec-

tive analysis of the current state of affairs will have to conclude that the general performance of the public sector institutions is not necessarily uniformly bad, and equally the achievements of all the private sector institutions are not necessarily uniformly good. Each case should be judged on its merits and demerits, and over a specific period of time. Accordingly, the high priests of the private sector institutions who claim that the private sector will solve all the water problems of the world, and the diehard social activists who claim that the private sector institutions have no role to play in water supply and wastewater management, are both wrong. Each project should be judged by its performance, which must be based on objective analyses of facts, and not by dogmas, personal biases and hidden agendas.

In addition, the discussions thus far have been almost exclusively framed in terms of the performance of the public and the private sector institutions that provide water supply and wastewater treatment services. There is, however, another model, where the water-related services are under the control of a public sector institution, but this institution could outsource specific activities which it considers the private sector can do more efficiently and cost-effectively than itself. These outsourced activities could include activities like information technology, meter-reading, bill preparation and collection, leak detection and repair, management of fleet of vehicles, etc.

The rest of this chapter analyses the recent performance of a Sri Lankan institution, which is under public control but has used private sector institutions extensively to carry out certain specific tasks.

2.3 Analysis of a Sri Lankan Institution

Objective and comprehensive analyses of water management institutions, be they public or private, are few and far between. The following analysis of a Sri Lankan institution was possible because of the support of the Japan Bank for International Cooperation and excellent collaboration from the Government of Sri Lanka.

The institutional analysis provided in this chapter is for the National Water Supply and Drainage Board (NWSDB) of Sri Lanka. The Board is responsible for water supply and drainage of the entire country. Because of time and financial constraints, it was not possible to analyse the institutional performance of the Board for the country as a whole, but rather for Colombo, and very specifically for the towns east and south of Colombo. Since the Board is a national institution, it is likely that its performance in managing water and wastewater in and around Colombo will not differ materially from the rest of the country.

It should be noted that in the final analysis, unless an institution is financially viable, its institutional sustainability and management efficiency on a long-term basis cannot be assured. Since access to water supply and sanitation is to be made universal in Sri Lanka, and wastewater treatments and disposal practices need to be very improved during the next decade, investments requirements for this sector will be extraordinarily high. Thus, economic sustainability of NWSDB as an institution is given special consideration in this chapter.

Since the Board functions on the basis of the national policies, and not individual project by project policies, it is necessary to analyse its current policies and performance in order to get a clear picture of its long-term sustainability. In other words, the macro policies and performance of the Board at the national level need to be examined. Future investments needed to meet the targets in the water supply and sanitation sectors are also an important consideration to assess the sustainability and proper functioning of the Board.

2.3.1 National Policy on Water Supply and Sanitation

According to the draft Sri Lankan National Water Policy (NWSDB 2002), provision of "safe drinking water and access to sanitation services is a fundamental element" for the social and economic development of the country. At present, 26.5 percent of national population receives piped water: access to sanitation services is significantly lower. The policy notes that "while coverage levels and service quality have improved markedly over the past decade, the need for water services has outstripped the government's ability to provide sufficient water and sanitation and ensure equitable access to the citizens throughout the country." This realistic policy goal is to be applauded, since achievements of this goal will go a long way to improve the existing quality of life of the people.

The draft policy further concludes that one of the important issues that have "outstripped the government's ability" is the need for funds for capital investments for new water supply and sewerage projects. In addition to these funding requirements, efficient operation and maintenance of the existing projects will also become an increasingly important issue in the coming years, especially as the number of new water supply and sewerage projects is likely to increase very significantly in the future. Thus, with shortages in the government funding, which already has become a serious constraint to assure future developments on a timely basis, it is essential that the customers of the Board at least pay for the full operation and maintenance costs of the existing and proposed water supply and sewerage projects, and if possible increasing percentages of the investment costs in the future. It will not be possible for the Sri Lankan Government to subsidize heavily the operation and maintenance costs of the Board, as well as the high capital investment costs of all its projects, and concurrently meet the universal service provision targets.

This, of course, is not a special situation that is applicable only for a country like Sri Lanka: nearly all other developing countries are facing very similar problems for their water supply and sanitation sectors, and also somewhat same type of economic constraints. For all developing countries, the time for free availability of the government-subsidized water and sanitation services has now become history. The consumers need to pay for these services based on the twin principles of economic efficiency and equity. However, different developing countries are taking different approaches to overcome the financial constraints, and also are adopting new management practices to improve their technical and economic performance, and concurrently provide better water and sanitation services to an increasing majority of their citizens. NWSDB needs to carefully assess these new approaches that are being tried in different countries to see to what extent they can be applied under the Sri Lankan conditions, perhaps with appropriate modifications that are considered necessary.

Because of these developments and the perceived constraints, the Government of Sri Lanka has initiated a reform programme for the water sector which includes, among other items, establishment of a regulatory commission for water supply and sewerage, and contracting private operators in selected regions to improve operational efficiency and to attract private sector investment finance. Both the sector reform programme, and the increasing involvement of the private sector, are likely to be important factors to improve the existing situations further in the future.

The draft National Water Policy (NWSDB 2002) proposes the following steps to enhance the economic sustainability of the institution. The proposed steps are the following:

- Water tariffs in the urban areas should be set at such a level that it should be possible to recover operating costs and depreciation, and should be gradually increased "to recover the full supply cost of providing services, including debt service and a reasonable rate of return".
- Cross-subsidies between commercial/industrial consumers and domestic consumers should be reduced to a reasonable level. However, what constitutes a "reasonable level", and how it can be estimated, are not specified in the policy. Currently there are also cross-subsidies between projects, which need to be progressively reduced, and perhaps eliminated for most cases, unless there are very special and specific justifications for maintaining them.
- Sewerage tariff that covers operation and maintenance costs shall be introduced, based on water consumption for the sewered areas, and also as and when sewerage services are introduced to these new areas.
- For low-income people, appropriate life-line tariff should be available to ensure the affordability of water of sufficient quantity and quality to satisfy both basic consumption and hygienic requirements. A process which identifies an appropriate life-line tariff for the poor, without subsidizing the rich, has not been easy to formulate, neither in Sri Lanka, nor in any other developing country. This aspect will require further intensified research in order to identify the various policy-options available to formulate an appropriate life-line tariff for

the poor under the Sri Lankan conditions, which should not subsidize the rich, who can pay for the services they receive.

- Operational efficiencies of the water supply systems should be improved, and the levels of non-revenue water in all projects should be significantly reduced from their current high levels. This is a very important consideration for Sri Lanka, both for existing and new projects.
- Water demand management programmes should be implemented to reduce the levels of current consumption. In addition to the proposed economic instruments, other possible alternatives should be considered (for example, education and communication) to increase the use of demand management practices very significantly.

All the above policy recommendations are unquestionably steps in the right direction to ensure the economic sustainability of the Board, and also to further improve its management practices. If and when these draft policies are fully implemented, they will go a long way to assure the economic sustainability of the individual water projects as well as that of the Board.

While the proposed policy is a step in the right direction, the main consideration for the future will not be the adequacy, or even the appropriateness and desirability of the policy, but its proper and timely implementation, especially as it will mean steady increases in the water and sewerage charges in the coming years in order to assure full cost recovery within a reasonable period of time. The charges that the customers will have to pay for water will be much higher than what they are at present, most probably by a very significant amount in the coming years, especially when sewerage and wastewater treatment services are introduced. This also means that special attention needs to be paid in terms of equity, so that the poor families are not unduly penalised by higher water and sewerage charges, which they may not be able to afford.

2.3.2 Water Tariffs

Since, under the existing conditions, the final decision to increase the water tariffs in Sri Lanka depends primarily on a political process, it is difficult to predict at present as to what extent the politicians will allow the tariffs to be increased to the level necessary for full cost recovery on a timely and regular basis. With the quality of the current senior management in the institution, it is highly likely that it will request the necessary tariff increases in a timely manner, as stipulated in the draft national water policy. However, the approval of the higher tariff structure is beyond the current mandate of the institution. Increase in water tariffs basically remains a political process, which means that the increases requested by the Board, may or may not be considered and approved by the politicians in a timely manner. The general experiences in developing countries have been that politicians generally prefer to decide to "go slow" to increase tariffs, especially before elections.

The political situation in Sri Lanka has unquestionably improved during the past decade, and the people and the politicians are now more aware than ever before that the consumers must pay for water, sewerage and wastewater treatment services in order to continue to receive a good service, and also ensure universal coverage in the country. The Government simply does not have enough funds to continue to invest heavily in these sectors and pay for their timely and proper operation and maintenance, and, in addition, provide a subsidised service to all the country's residents. Thus, unless some radical improvements are made in the autonomy and the responsibilities of NWSDB as an institution, one possibility still remains that the political process may not allow the water tariffs to be increased regularly, at appropriate intervals, to the level necessary for full cost recovery. There is a good possibility that prices will be allowed to increase, but not all the time, and not to the level necessary, and also not exactly promptly when they are requested by the Board. Such developments could be noted in the recent past. Furthermore, the multilateral and bilateral donor agencies have also becoming increasingly aware in recent years of the importance of practicing water pricing in terms of cost recovery and water conservation. Most of these agencies are now requiring, as a part of their loan agreements, that the Government initiates and ensures water tariff increases to more appropriate economic levels.

The urban domestic water tariffs in Sri Lanka are still heavily subsidised by the Government, and by cross-subsidies between consumers, and also between various projects. These various subsidies, however, have been steadily declining, especially during the past ten years. The current Government subsidies include:

- 50 percent subsidy for the foreign loan components; and
- 100 percent subsidy for the Sri Lankan Government contributions that is available in the local funds.

At present, the Board is required to pay back to the Government only 50 percent of the foreign loans, at an annual interest of 10 percent, over a period of 24 years, with another two years of grace period, if considered necessary. Furthermore, the Board returns the funds to the Government in Sri Lankan rupees: the Government assumes the entire foreign exchange risk, which could be quite substantial under many conditions.

The revenue collection for the urban water supply in Sri Lanka started in 1982. Considerable progress has been made since that time, especially during the post-1995 period. There have been continuous improvements in the methods used to set the tariff structures, the levels of the tariffs, and the collection of the revenues outstanding from the consumers.

2.3.3 Other Economic Considerations

Because of the increasing investments in the water sector, the total debt service costs of the Board have steadily increased from Rs. 208 million in 1997, to Rs. 521 million in 1999, Rs. 684 million in 2000, Rs. 704 million in 2001, and Rs. 854 million in 2002. With heavy investments necessary for the next 10 years to meet the targets for the urban water supply and sewerage services, as stipulated in the draft water sector policy, the level of the debt service for the Board will increase steadily, and most likely very appreciably, in the foreseeable future. Because of these increasing debt service payments, fiscal prudence and higher institutional efficiency will become two very important requirements for the institutional sustainability of the Board in the future.

In order to meet this increasing debt service costs, as well as higher operation and maintenance costs, and reduced cross-subsidies between the consumers and the projects, the water tariffs will have to be increased steadily and regularly over the next decade and beyond, and also NWSDB has to become an increasingly more efficient institution so that its costs of providing the necessary services can be reduced to as low a level as possible, and also as soon as feasible.

The income of the Board from the sale of water at present depends not only on the prevailing tariffs but also on the number of connections, amount of water used by the various categories of users, billing accuracy, and the bill collection efficiency of the Board. All these issues will be considered one by one next.

The average monthly consumption of water for 2000 and 2001, and the average household monthly bills for the same years are shown in Table 2.1. These factors also have bearings on the total income of the Board, and thus ultimately on the economic sustainability of this institution.

Table 2.1. Average monthly consumption and bill per household, 2000-2001

	Monthly consumption (m^3)		Monthly bill (Rs.)	
Areas	2000	2001	2000	2001
Greater Colombo	22.4	21.95	196.62	235.17
Regions	17.7	16.5	113.02	125.67
Average, Sri Lanka	19.67	19.15	154.27	178.95

Source: NWSDB, non-published information

If a longer-term trend is analysed in terms of average monthly household consumptions, the trend from the year 1996 has been one of a regularly declining one.

For example, in 1996, an average household consumed 25.2 m^3 of water per month in Greater Colombo, and 18.8 m^3 in the regions. These consumption levels have steadily declined to 21.6 m^3 and 16.3 m^3, respectively, by the 2002 (the figures are for up to September for both the years concerned). This indicates that there was a net reduction of household consumption by 14.28 percent in Greater Colombo, and 13.3 percent in the regions, within this seven-year period. Further reduction in consumption is a current objective of the Board in order that more people can share the same quantity of available water.

Similar trends can be observed at the per capita level. In 1995, per capita consumption in Colombo was around 200 litres. The consumption level came down to about 140 litres by 2002, primarily through water pricing. If this can be further reduced to 100 litres, a level that is feasible, the current total demand of 246,000 m^3/day, could be reduced to 185,000 m^3/day, a reduction of about 60,000 m^3/day. Coincidentally, The Phase 1 of the proposed Kalu Ganga project will deliver an identical amount of water, that is, 60,000 m^3/day, but at a capital investment cost of $84 million U.S. dollars. Thus, soft options, like demand management, promotion of water conservation, etc., need increasingly more attention from the institution to meet the identical social goals. Many times, some of the soft options can be implemented at cheaper economic and environmental costs, and within a much shorter time-scale, compared to hard options, like construction of new water projects.

The average water consumption in the Greater Colombo area declined by about 2 percent, and in the regions by 6.8 percent, over the period of 2000-2001. However, during the same period, the average household bill increased by about 20 percent in Colombo, and by 11.2 percent in the regions. While this is to be expected in terms of overall direction because of increasing focus on demand management practices by the Board, especially through higher tariffs, an interesting anomaly could be noted. Even though the increase in the average household bill in Colombo was higher than the regions by about 46 percent, the decline in water consumption in the regions during the corresponding period was significantly higher than in Colombo. This anomaly can perhaps be explained by the facts that the average income of the households in Colombo is higher than in the regions, and the average household water bill, even after the increase of the tariffs, is still quite low, about 1 percent of the average household income. Thus, unless the average monthly expenditures on water become a higher proportion of the monthly average household income, their effectiveness in terms of demand management is likely to be limited. Elasticity of water demand is likely to be significant at somewhat higher tariffs.

In terms of water sold and income to the Board, some interesting conclusions can be drawn based on the information available from Table 2.2, which shows the data for 2001. Slightly over 60 percent of all water produced was sold to the domestic consumers, but it accounted for about 36 percent of the total revenue of the

Board from the total water sales. In contrast, the various government institutions consumed around 11.33 percent of the water sold, but accounted for 25.18 percent of the revenue, and the commercial units consumed 9.15 percent of water sold, and paid 21.05 percent of the revenue. This indicates that the domestic consumers are still being subsidized to a very significant degree. According to the draft water policy, these types of cross-subsidies are to be reduced to "reasonable" levels, which will mean that the domestic consumers will have to pay progressively higher charges for water and sewer services in the future.

Table 2.2. Quantities of water sold and revenues by consumer categories, 2001

Categories	Quantities sold (1000 m^3)	Percentages	Revenues (million Rs.)	Percentages
Domestic	134,691	60.39	1,258.50	35.97
Government	25,276	11.33	880.86	25.18
Commercial	20,410	9.15	736.45	21.05
Standposts	14,129	6.34	83.84	2.40

Source: NWSDB, non-published reports

Another important factor that has an impact on the total revenue is the efficiency of the collection of the bills by the Board. Table 2.3 shows the actual amount collected against the water bills for the period 1993-2001. In terms of bill collection, it has to be admitted that the performance of the Board has been exemplary during the past decade, and this high level of revenue collection is seldom matched in other developing countries. The total bill collection efficiency varied from a low of 89 percent to a high of 99 percent during these nine years. The bill collection efficiencies of the Board for the 1993-2001 period are shown in Figure 2.1.

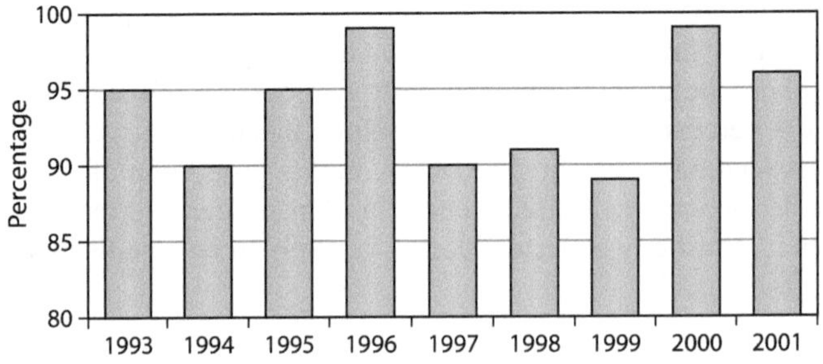

Fig. 2.1. Bill collection efficiencies of the Board, 1993-2001

One of the reasons as to why the Board has an excellent bill recovery record is because it has an elaborate, well-established and transparent process. Consumers are well aware of this process (Figure 2.2), and they are also aware that if the bill continues to be unpaid, the water connection will be cut-off. In addition to the disruptions and inconvenience it will invariably cause to the various household activities, there is an economic penalty in terms of an additional reconnection charge. The connection charges, after disconnection, are Rs. 500 for the households, and Rs. 1500 for all others.

The Board has recently proposed an incentive-disincentive system for prompt bill payments. A 4 percent discount on the bill will be allowed for advance payments of the bills. A 2 percent discount in the next month's bill will be made, if the bill is paid within four days of its receipt. In addition, it has been proposed that a deterrent charge of Rs. 100 will be levied from 2003 for non-payment. These progressive developments are likely to ensure that the Board will continue to have a good bill recovery record

2.3.4 Non-Revenue Water

A very important economic loss to the Board has been due to the extent of non-revenue water (NRW), which is defined by the difference between the quantity of water produced and the quantity billed for. At present NWSDB is billing for only about 65 percent of the total amount of water it produces in the Greater Colombo area. In other words, the balance of 35 percent of the water produced unfortunately does not generate any revenue. This means that if the entire non-revenue water loss can be completely eliminated (this of course is not possible), the Board's income will increase by more than 50 percent at the current prevailing water tariffs.

If Colombo City is considered, NRW accounts for more than half the quantity of water that is billed for. The composition of NRW for NWSDB as a whole and Colombo City is shown in Table 2.4. The NRW rates are high, but such high rates of NRW are not exactly uncommon in the major urban centres of developing countries, which generally range from 40 to 60 percent. The NRW component in the Colombo City is thus at the higher end of this spectrum. In other words, these losses mean that in Colombo City, NWSDB loses slightly more than half of its possible revenue.

Table 2.3. Percentage of Collection Against Revenue (Rs. 1000), 1993-2001

YEARS	REVENUE			COLLECTION			% COLLECTION AGAINST REVENUE		
	Domestic	Non domestic	Total	Domestic	Non domestic	Total	Domestic	Non domestic	Total
1993	260,389	890,765	1,151,154	255,582	836,078	1,091,660	98	94	95
1994	309,007	1,020,738	1,329,745	259,624	951,423	1,211,047	84	93	91
1995	356,694	1,167,558	1,524,252	336,643	1,110,180	1,446,823	94	95	95
1996	373,009	1,191,356	1,564,365	348,198	1,195,241	1,523,439	93	100	99
1997	545,817	1,484,508	2,030,325	451,523	1,367,074	1,818,597	83	92	90
1998	714,257	1,760,861	2,475,188	629,901	1,626,542	2,256,443	88	92	91
1999	802,232	1,912,136	2,714,368	698,984	1,726,878	2,425,862	87	90	89
2000	963,925	2,042,556	3,006,481	1,017,500	1,962,121	2,979,621	106	96	99
2001	1,284,627	2,213,778	3,498,405	1,234,486	2,123,119	3,357,605	96	96	96

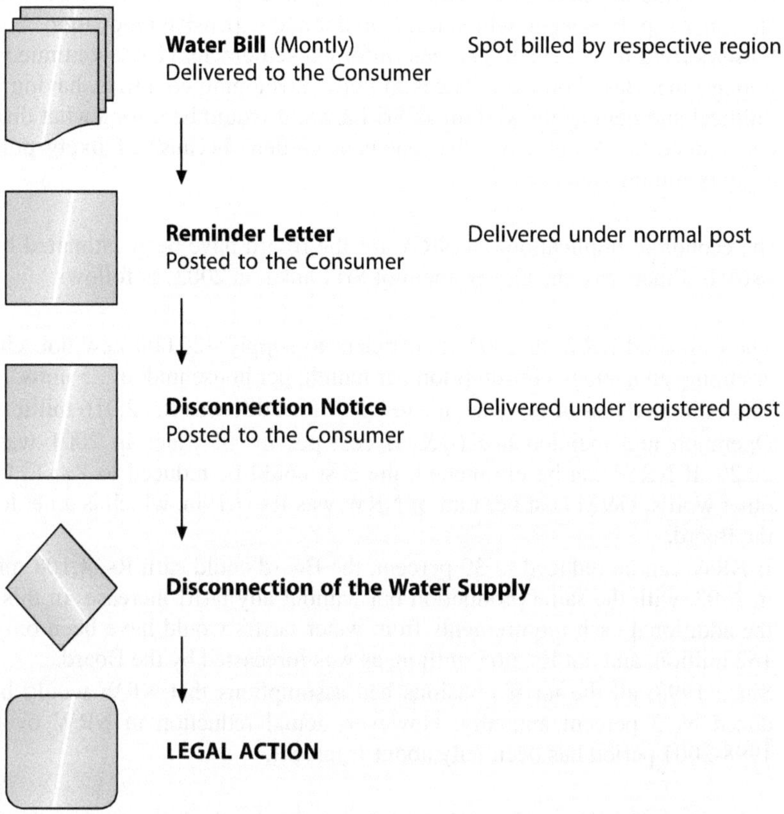

Fig. 2.2 Bill Recovery Process

Table 2.4. Non-revenue water for NWSDB and Colombo City as percentages of water produced

Type of losses	Overall NWSDB	Colombo City
Leakages	23	25
Tenement gardens, standposts	4	15
Illegal connections	4	8
Metering errors	4	5
Total losses	35	53

Source NWSDB, non-published information

The high rate of NRW in Colombo City is primarily due to the fact that it has an old water supply system, which needs major and expensive renovation. Equally high losses from the tenement gardens (informal settlements) have continued for a very long time. Based on experiences in other developing countries, having similar political and democratic system as Sri Lanka, it would be a somewhat difficult task to reduce the losses from the tenement gardens because of likely political constraints and interference.

The economic implications of NRW for the Board have been estimated by the Ministry of Finance of the Government of Sri Lanka, in 2002, as follows:

- The estimated NRW in 2002 is sufficient to supply 420,000 new households, assuming an average consumption per month, per household, of 25 units.
- The estimated loss of revenue due to NRW for 2001 was Rs. 2,016 million.
- Operation and maintenance (O&M) cost per m^3 of water in 2001 was Rs. 20.20. If NRW can be eliminated, the cost could be reduced to Rs. 13.01. In other words, O&M cost per unit of NRW was Rs. 7.19m, which is a net loss to the Board.
- If NRW can be reduced to 30 percent, the Board could earn Rs. 4,108 million in 2002, with the same production but without any tariff increase. In this case the additional cash requirements from water tariffs would have been only Rs. 162 million, and not Rs. 665 million, as was forecasted by the Board.
- Since 1998, all the tariff revisions had assumptions that NRW would be reduced by 3 percent annually. However, actual reduction in NRW over the 1998-2001 period has been only about 1 percent.

Reduction of NRW is an important consideration for the following reasons:

- It is by far the cheapest and the quickest method to increase the availability of water supply through "softer" methods by reducing the NRW component of water produced, compared to the "harder" options of construction of new water supply development projects to bring additional water from new sources.
- Reduction of NRW will immediately increase the revenue of NWSDB.
- More consumers can be provided with house connections from the saving of water due to the reduction in the NRW component. If urgent actions are taken to reduce the NRW component, new consumers can begin to receive benefits within a short period of about 10-12 months, compared to 6-15 years for planning and implementation of additional construction projects.

It appears that out of the 85,000 connections in Colombo City (this accounts for approximately 25 percent of total connections in the Greater Colombo area), nearly 30,000 households receive estimated bills, and at least another 20,000 water-meters are defective in this area.

Thus, installation of some 50,000 new meters alone on an urgent basis, will probably reduce the extent of NRW component in Colombo City by 15-17 percent within a limited period of 1-2 years. People in the poorer household could be encouraged to take house connections, which could be metered, but with more attractive economic options made available for the payment of the connection charges. These "softer" options to reduce NRW do not consider water losses due to leakages, which need to be tackled as well. Leak reduction will be more expensive and will take a somewhat longer time, compared to the above option.

It should be noted that the current NRW component in Singapore is only 6 percent and, in Tokyo, about 7 percent. Thus, there is a significant scope for the reduction in NRW in the future.

2.3.5 Policy Options for Improving Economic Sustainability

There are many policy options available to improve the economic sustainability of providing water services. These options are not mutually exclusive. Among the options that could be considered are the following:

Make the Board increasingly more efficient in terms of staffing

There is no question that the Board was heavily overstaffed a decade ago. For example, in 1991 the ratio of staff members in the Board per thousand connections was extremely high, at 33. The ratio has declined steadily, and by 2002, it was 11, which is only 1/3rd of what it was only 12 years ago. This is shown in Figure 2.3. This ratio is still high, when compared to the situations in other countries:

Singapore: 2 employees per 1,000 connections
Developed countries: mean 2.1-2.2 employees per 1000 connections
Thailand: 5 employees per 1,000 connections.

According to NWSDB, the water tariff for the year 2002 was formulated with the following assumptions:

- Number of new connections: 58,850;
- Average number of persons per household: 5;
- Replacement of 10 percent of metres;
- Reduction in the number of employees per 1000 connections: 0.5;
- Reduction in non-revenue water component: 3 percent;
- Estimated collection of bills: 98 percent; and
- 20 percent of capital budget (local component) to be recovered as overheads.

The situation can be viewed in another way. In 2002, the personnel cost as percentage of the total operating cost was approximately 43 percent for the Board. In the developed world the corresponding figure is around 30 percent. Thus, there is still considerable scope to reduce staffing situation of the Board. Even though the progress in reducing over-staffing in the Board has been excellent during the past decade, much progress still has to be made in the coming years.

Cross-subsidies

Cross-subsidies between the different consumer categories and between the different projects need to be reduced so far. The cross-subsidies between the different consumer categories are being progressively reduced. For example, cross-subsidy between domestic and non-domestic customers have already been reduced to 1:5.06. However, similar reduction in cross-subsidies is not taking place between projects.

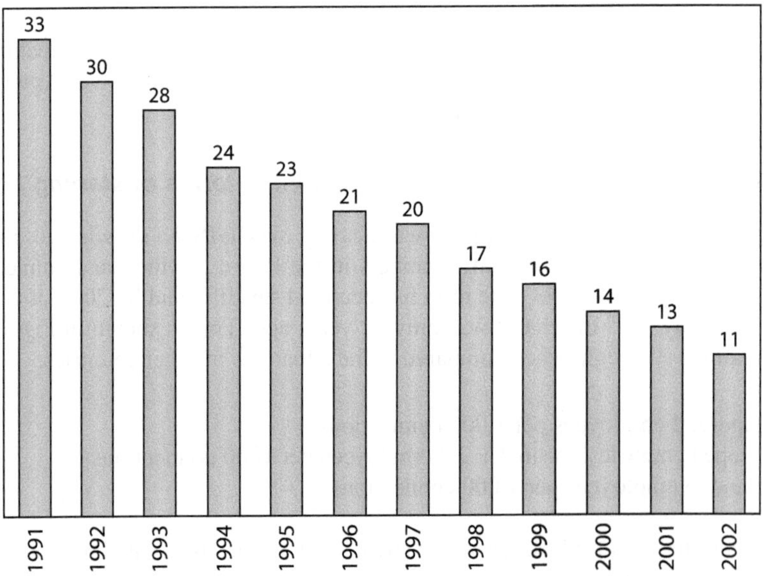

Fig. 2.3. Staff members per 1,000 connections

Calculations made by the Sri Lankan Ministry of Finance (2002) indicate that the operation and the maintenance costs of all projects are expected to be covered by the tariff revision of 2002. However, recovery of scheme-wise operation and maintenance cost is still a long way off. With the earlier tariff that was valid in 2001, 137 projects were considered to be loss-making, and 44 were profitable. Even with the new revised tariffs, a vast majority of the loss-making schemes will still remain loss-making. In other words, some water projects pay more than their fair share of the O&M costs, whereas the majority of the projects pay less than their actual O&M costs.

The Ministry of Finance (2002) further notes that "during the past few years several projects were approved by the Cabinet subject to the implementation of scheme-wise tariff in those particular schemes. But they were not implemented".

Private sector involvement

The issue of institutional sustainability cannot be answered by water tariffs and current revenues only: other factors need to be considered as well. Among these other factors are institutional efficiencies and non-revenue water. Logically and rationally, consumers should not be expected to pay higher prices for the inefficiencies and shortcomings of any institution. Accordingly, if full cost recovery for water and sanitation services is to be the final objective, it is essential that all water-related services be made as efficient and as rational as possible, so that the water users may be prepared to pay higher prices that are consistent with a lean operation.

In this connection, it may be desirable to consider the possibility of outsourcing many of the activities that are currently being done by the Board. It should be noted that the Board has been increasingly outsourcing more and more activities in recent years. It is thus not a new phenomenon for the Board, but much more remains to be done.

Probably the most important group of people in the Board that the water users are in regular contact with is the meter-readers. The current practices of the Board, where meter-readers are given specific areas to cover consistently over a very long period of time, have not produced good results in other developing countries. Such territorial control by meter-readers often has resulted in increases in their unauthorized incomes, since for some staff members, personal interest may override the overall public interest. In order to counteract corruption, the meter-readers need to be rotated very frequently in terms of the areas they cover, so as to ensure the potential of unauthorized incomes is significantly reduced, or even totally eliminated. For example, such very frequent rotations are practiced in Singapore, where corruption for meter reading has been virtually reduced to zero.

Based on experiences in other South Asian countries, it is estimated that the Board's income could probably be increased within the range of Rs. 225-Rs. 350 million by proper, accurate and efficient meter-reading. In addition, such transformation could be made within a very short period of only about 1-2 months.

It would, however, most likely to be a very difficult political process for the Board to change the current meter-reading practices, and change the territories of the meter-readers, who are highly organised and are politically well-connected.

The meter-readers, in all probability, will resist such a move very strongly because of their own personal vested interests.

Accordingly, one realistic solution to improve the situation could be to outsource the meter-reading to the private sector. Such a step is likely to improve the present situation very significantly. The current systemic corruption in water-reading is likely to be reduced substantially. In addition, it will mean that the overall cost of meter-reading will be reduced significantly because not only the number of meter-readers will be less, but also since the private sector is unlikely to pay the existing salaries and fringe benefits that they currently receive.

The Board could provide an incentive to the private sector that they will receive a percentage of additional revenue that the may be able to generate from the existing consumers. If these proposed steps are applied, it is likely that the net benefit to the Board, in terms of additional revenue and reduced cost of the provision of meter services, will be in excess of Rs. 400 million per year, a not inconsiderable sum.

There are many other activities that the Board could consider outsourcing to the private sector. Among these possibilities are:

- Leakage detection and repair:
- bill collection;
- new water connections; and
- vehicle maintenance and fleet management

All these activities, both individually and collectively, are likely to contribute to steady advances to the institutional sustainability of the institutions.

2.4 Concluding Remarks

An analysis of the institutional performance of the Board indicates that it has made tremendous advances over the past 15 years. All the performance indicators analysed have improved significantly. While these are commendable and are undoubtedly steps in the right direction, much more remains to be done in the coming years.

The magnitudes and complexities of water planning and management practices have changed radically during the post-1990 period, so much so that many of the approaches, techniques and solutions that were used extensively earlier may not be appropriate any longer, or may not be acceptable in social, political and institutional terms. Issue like decentralization, public participation, private sector involvement, water pricing, water rights, social and environmental consideration, transparent governance, globalization, information and communication revolution, and increasing emphasis on poverty alleviation were not explicitly and/or comprehensively considered earlier.

Water management institutions must respond to these changes and new requirements. Institutional arrangements, even when they have worked perfectly well during the pre-1995 era, must change radically if they are to meet the challenges and rapidly changing boundary conditions of the post-2000 period. The extent and magnitude of the changes during the next two decades are likely to be unprecedented. To paraphrase Lewis Carroll, the water institutions of the future will have to run faster just to stay in the same place.

2.5 References

Al Baz I, Biswas AK (2003) Special thematic issue: public private partnership in the Middle East and North Africa. International Journal of Water Resources Development 19: 115-232

Ministry of Finance, April 25, 2002, Note on water tariff revision 2002, to Ministry of Housing and Plantation Infrastructure, pp 1-3

NWSDB (2002) National policy on water supply and sanitation. National Water Supply and Drainage Board, Ministry of Housing and Plantation Infrastructure, Colombo, Sri Lanka

Tortajada C, Biswas AK (2004) Precio del agua y participación pública-privada en el sector hidráulico (in Spanish). Miguel Ángel Porrua, México, DF

World Commission for Water in the 21st Century (2000) A Water Secure World: Vision for Water, Life and the Environment, World Water Council, Marseille

Water Institutions in India: Structure, Performance, and Change

R. Maria Saleth

3.1 Introduction

With increasing water scarcity and frequent occurrence of water-related conflicts at macro and micro levels, the institutional arrangements governing water resource development, allocation, and management are receiving increasing policy attention worldwide. The water institutional arrangements assume, however, an enhanced significance in India as the country is moving fast towards the ultimate limit of its utilizable water resource potential. Although India is using only about 57 percent of its total water resource potential at present, it is already using about 66 percent of its irrigation potential [Government of India (GOI) 2000]. While water demand is increasing fast with a growing population and an expanding economy, further development of water resources is seriously constrained by investment bottlenecks, environmental concerns, and political and legal snags inherent in inter-regional water transfers. As the gap between projected demand and potential supply is likely to grow further, the physical scarcity of water, which is already visible in a few regions and cities, is expected to assume a national proportion in the not-too-distant future.

Unfortunately, most water institutions—being developed in an era of water surplus, especially during the colonial period—are becoming increasingly less relevant to address the water challenges as the country is entering in an era of absolute water scarcity. While Indian water institutions are undergoing significant changes in recent years, these changes fall far short of the new and emerging institutional requirements of the water sector. To see the magnitude and consequences of this reform gap, we need to first review the water institutional structure and then, attempt an evaluation of its efficacy and performance. Both the review of water institutional structure and the evaluation of its performance can be more illuminating when they are based on a conceptual and analytical framework based on the institutional decomposition approach. Similarly, the nature and direction of recent institutional changes as well as the factors behind these changes can be grasped better within a theoretical framework based on institutional transaction cost approach that accounts for both the economic and political transaction costs.

3.2 Objectives and Scope

The overall objective of this chapter is to review the historical evolution, present structure, functional performance, ongoing changes, and future direction of the water institutional arrangements in the particular context of India against a brief description of the water institutional environment within which the water institutional structure is embedded. This chapter attempts to realize its overall objective by addressing the following specific objectives:

(a) outline the analytical framework and the theoretical approach being used to review water institutional structure and explain its evolution and change;
(b) provide a brief description of the water institutional environment as characterized by the historical, socio-economic, political, and physical factors;
(c) review the evolution, structure, and performance of water institutional arrangements, focusing first on the macro/formal institutional arrangements and then, on their micro/informal counterparts;
(d) assess the performance of water institutions using few objective criteria;
(e) evaluate the recent institutional changes within an institutional transaction cost framework and identify the forces—both endogenous and exogenous to the water sector—that determine the depth and direction of water institutional changes; and
(f) conclude by identifying some major implications for theory and policy in the realm of water institutional reforms.

As to its focus and scope, this chapter is more eclectic rather than exhaustive in terms of its coverage of the legal, policy, and administrative or organizational aspects governing the water sector. The emphasis will be on the most important aspects of water institutions that are receiving considerable attention in the current debate on water sector reforms both in India and elsewhere. Such a selective coverage is necessary partly due to the need for a more focused analysis and partly due to the continental nature of India having a wider regional diversity in terms of both water institutional arrangements and water sector features. Nevertheless, as much as possible, region-specific aspects of both water institutional arrangements as well as water sector features will be highlighted in all relevant contexts. While informal institutional arrangements operating at the micro level will receive attention, the major focus will be on the formal institutional arrangements that are operating at the national and regional levels. This is because the formal macro-level institutions are more amenable for purposive reforms as compared to their informal and micro-level counterparts.

3.3 Analytical Framework

Institutions are entities defined by a configuration of legal, policy, and organizational rules, conventions, and practices that are structurally linked and operationally embedded within a well-specified environment. A simple analytical decomposition can enable us to have a better understanding of their nature and features. From a very broad perspective, institutions can be decomposed by distinguishing the institutional structure (or, governance structure) from its institutional environment (or, governance framework) (Williamson 1975; North 1990a). While the institutional environment is characterized by the overall cultural, historic, socio-economic, and political milieu of a country or region, the institutional structure is defined by the interactive effects of legal, policy, and organizational or administrative components and their constituent aspects. Since the institutional structure is embedded within the institutional environment, the evolution of the former is invariably conditioned by changes in the latter. This does not mean that there is only one-way flow of effects between the two, as changing institutional structure also influences the institutional environment or governance framework.

Like all other institutions, water institutions also have their own institutional structure, as well as, their institutional environment (Saleth and Dinar 1999, 2003). The water institutional environment is characterized not only by the factors determining the overall institutional environment but also by those related to water resource conditions as well as to other water-related sectors such as agriculture, environment and forest, and urban development. The water institutional structure is defined interactively by three institutional components, i.e., water law, water policy, and water administration (or, water-related organizations). The institutional components cover not only the formal and macro-level arrangements but also the informal and micro-level arrangements such as those reflected in local customs, conventions, and informal contracts.

The formal *institutional components* (to some extent, their informal counterparts) can also be decomposed further to highlight some of their major *institutional aspects* (Saleth and Dinar 1999, 2003). For instance, water law can be decomposed to highlight: (a) inter-governmental responsibility, (b) water rights, and (c) accountability provisions and mechanisms. Similarly, water policy can be decomposed to shed light on: (a) project selection criteria, (b) pricing and cost recovery, and (c) user and private sector participation policy. In the same way, the organizational dimension of water sector can be decomposed to focus on: (a) organizational structure and the relative role of government layers, (b) financing and management, (c) regulatory mechanisms, and (d) conflict resolution arrangements. It is this institutional decomposition that we will use here as the analytical framework for the review and description of the water institutional structure in India. Two critical features of this analytical framework can be noted. First, although it is not that detailed and exhaustive, it still captures most of the institutional issues that are currently receiving wider attention both in national and global policy debates. Second, the decomposed institutional components and their constituent institutional aspects are treated as independent entities just for analytical convenience. In reality, the institutional aspects are functionally nested and interlinked both within and across the institutional components.[1]

The theoretical framework that we use here for explaining water institutional evolution and change is based on a simple representation of the institutional transaction cost theory (Coase 1937; Williamson 1975; North 1990b) as attempted by Saleth and Dinar (2000 and 2003). In short, this framework identifies first all the major factors affecting water institutions and then, traces their effects on the transaction and opportunity costs of institutional reform. While the factors influencing water institutions are many and diverse, for analytical convenience, they can be grouped into endogenous factors that are internal to water sector and exogenous factors that are outside the strict confines of both water institution and water sector. The endogenous factors include water scarcity, water conflicts, financial and physical deterioration of water infrastructure, and operational inefficiency of water institutions. The exogenous factors include economic development, demographic growth, technical progress, economic and political reforms, international commitments and pressures, changing social values and ethos, and natural phenomena such as floods and droughts.[2]

Although it is difficult to isolate the individual effects of these exogenous and endogenous factors, it is still possible to track them, especially by conceptualizing their effects either in terms of the transaction costs or in terms of the opportunity costs. The transaction costs cover both the real and monetary costs of instituting and changing the regulatory, monitoring, and enforcement mechanisms related to water development, allocation, utilization, and management. The opportunity costs, on the other hand, cover both the real and economic value of opportunities foregone (i.e., the net social costs of inaction or '*status quo*'). The institutional transaction cost theory asserts that institutional change occurs whenever the opportunity costs exceed the transaction costs so as to give the much needed political economy thrust for reform.

The opportunity and transactions costs of institutional changes are not static but change continuously due to changes in factors both endogenous and exogenous to

---

[1] For instance, within water law component, the aspects of water rights, conflict resolution, and accountability are interrelated as do the aspects of pricing, cost recovery, and user and private participation within water policy component. Since water policy is a political translation of water law (or, water law is a legal translation of the policies that have attained political consensus or withstood the test of time) and water administration is the implementation arm for water law and water policy, these three institutional components are nested and linked both organically and operationally.

[2] Notice that it is these exogenous factors that together, in fact, characterize the overall institutional environment for the water institutional structure. As such, our analytical framework captures also the transaction cost implications of changing institutional environment that is characterized by both economic and non-economic factors.

the water sector. For instance, as water scarcity becomes acute due to economic development and population growth, the real and economic costs of inappropriate water institutions tend to rise. Similarly, the economic reforms magnify the fiscal implications whereas natural calamities such as droughts and floods aggravate the political implications of the opportunity costs of institutional reforms. Political reforms involving nation-wide institutional changes, on the other hand, reduce the transaction costs of water sector reforms directly because the institutional changes within water sector form only a small part of the overall reform process. Likewise, technical progress can also reduce the transaction costs of institutional changes. As the exogenous factors tend to magnify the opportunity costs of water crisis and reduce the transaction costs of water sector reforms, they often provide a powerful economic urge and political thrust for water institutional changes.

Institutional change is not a one-time event but rather a continuum involving gradual changes over time in response to the changing dynamics and relative magnitude and distribution of the reform costs and benefits. As the reforms initiated in earlier stages brighten the prospects for downstream reforms, there are intricate linkages between the transaction costs of subsequent reforms and the opportunity costs of earlier reforms.[3] Similarly, since the institutional changes within water sector derive considerable synergy from exogenous factors that reflect changes elsewhere in the economy, the transaction costs of water sector reforms can also decline due to scale economies in institutional change.[4] The opportunity and transaction costs implications of these dynamic and internal features of the institutional linkages and reform sequencing are an important and inherent part of the institutional transaction cost theory (Saleth and Dinar 2003).

The institutional transaction cost theory, as outlined above, captures not only the role of factors both within and outside the water sector but also the strategic significance of certain dynamic aspects of institutional change such as institutional linkages and scale economies (Saleth and Dinar 2003). Since it provides a unified framework to track and account for the effects of various factors affecting institutional changes in the water sector, this theory can indeed be used to explain both country-specific and cross-country variations in the nature and direction of water institutional changes. While the set of factors affecting water institutional change does not vary much across countries, their relative role and significance in the opportunity cost-transaction cost calculus can vary considerably across countries. It

[3] For instance, with the establishment of a transferable water rights system, the creation of other institutional aspects such as the conflict resolution mechanisms and water markets becomes easier due to the linkages that the transactions costs of the latter two institutional aspects have with those of the former.

[4] The scale economies in transaction costs emerge from the fact that the cost of transacting water institutional changes is lower when water sector reform forms part of an overall country-wide economic reform (e.g., China) and political reconstruction (e.g., South Africa) than otherwise (Saleth and Dinar 2000).

is the contextual nature of these variations that, in fact, explains why countries differ in terms of the extent and depth of water institutional reform observed both in the past and at present (Saleth and Dinar 2000). It is this framework that we use here to explain water institutional changes in India.

3.4 Water Institutional Environment: An Overview

The institutional environment of water institutions in India can be characterized in terms of a synoptic overview of the country's history, constitutional framework, socio-economic conditions, political arrangements, and finally, the physical setting of the water sector itself. India has diverse cultural traditions and a rich history. Its present administrative and judicial systems have evolved through a long process involving a strong central Asian and Persian influence during the Mogul period and British influence during the colonial period. It is, however, the British who improved and perfected these diverse systems into a relatively uniform and a somewhat centralized administrative and judicial system. With the consolidation of centralized bureaucracy and the spread of market and commercialisation, local and community-centred institutions have lost their relevance and gradually disappeared. The British influence can also be seen in the Indian political system characterized by multi-party democracy based on the parliamentary system. India has a federal form of government operating within a modern Constitution that demarcates the responsibilities of the central, state, and local governments.[5]

India covers an area of 3.29 million km^2 and has a population of over a billion. Despite the constraints and burden imposed by heavy demographic pressure, Indian economy is able to grow annually around 5 percent with a per capita income of US$ 450.[6] Comparing the pre-independence situation with the present, India has shown a remarkable economic performance with the achievement of food self-sufficiency since the early 1970s and also self-reliance on most industrial products since the early 1980s. But, India is still largely a rural economy as two-thirds of the population still live in rural areas, though the rural areas themselves have undergone remarkable transformation in recent years, thanks to the expansion of transport and communication facilities. Rural poverty, which used to be as high as 56 percent during 1973-74, has declined to 39 percent during 1987-88 (Planning Commission 1993:37-40). Currently, this figure hovers around 20 percent. Although the share of agriculture in the gross domestic product has declined now to 30 percent, this sector is critical both as a major source of employment (70 percent

[5] Although the central government is strong as per the Constitution, the state governments are becoming politically powerful in recent years due to the growing influence of regional parties both in state assemblies and in parliament.

[6] When evaluated in terms of purchasing power parity, this per capita income is equivalent to about $2,150 (World Bank 2000).

share in total employment) and as a key predicator of the pace and tempo of the overall macro economic performance itself.

Understandably, the water economy plays a critical role in the overall structure of the Indian economy. Being a vast and monsoon-dependent country, water resource potential displays severe spatial and temporal variations.[7] From an aggregate perspective, the total water resource potential of the country is estimated to be about 1,953 billion cubic meters (bcum), of which only 1,122 bcum can be utilized under current economic and technological conditions. However, the actually developed resource at present is only about 644 bcum representing 57 percent of the utilizable potential. But, the total water requirement of the country is projected to be in the range of 694-710 bcum by the year 2010, 784-850 bcum by the year 2025, and 973-1,180 bcum by the year 2050 (GOI 2000). In view of such an increasing supply-demand gap, especially in the face of economic growth and demographic expansion, the per capita water availability, which has declined from about 5,277 cubic meters (cum) in 1955 to 1970 cum at present, is likely to get depressed further. Although non-irrigation demand is likely to quadruple, the essentially rural and agricultural basis of the Indian economy will continue to orient the water sector essentially towards its irrigation sub-sector.

One key feature of the water sector with considerable institutional implications is the administrative demarcation of different water sub-sectors. While the canal irrigation sector is developed and managed by public agencies, the groundwater irrigation is developed and managed by millions of independent farmers.[8] Groundwater irrigation is made possible by 9.8 million electric and 4.4 million diesel pumpsets that are fitted with some 5 million shallow/deep tubewells—located mainly in the Indo-Gangetic and deltaic regions—and 10 million dugwells—spread essentially in the hardrock region covering western and peninsular India. Besides, there are about 0.17 million surface water-based lifts located mostly in the deltaic regions of West Bengal and Orissa. On the other hand, water supply in urban and rural areas is largely managed by semi-autonomous water supply undertakings, municipalities, and local governments. Finally, institutions related to other resources involved in water resource development and utilization as well as those related to general economic and sectoral management also form part of the water institutional environment. These include the land, forest, and agriculture-related institutions (e.g., land tenure and tenancy, forest and environment acts, agricultural pricing policies, and trade policies and international agreements).

---

[7] For instance, the average annual precipitation varies from 130 millimeters (mm) in Rajasthan desert to 11,000 mm—the world's highest rainfall—in the Assam Mountains. Notably, three-fourths of the rainfall in India is received in just four months during June-September.

[8] However, groundwater development is supported by government both directly in the form of public or state tubewells as well as indirectly in the form of credit support and massive investment in rural electrification programmes.

3.5 Water Institutional Structure: A Macro Perspective

For a more focused treatment, the review of the structure of formal and macro-level water institutions is organized within the analytical framework based on institutional decomposition. This framework distinguishes three institutional components, i.e., water law, water policy, and water administration, and highlights a few key institutional aspects under each of these three components. As noted already, the ensuing review is focused only on some of the key institutional aspects under each of these three institutional components.

3.5.1 Water Law

Water law assumes a central place in the functioning of water institutions as it gives the full legal backing to water policy as well as provides the operational framework and enforcing power for water administration including its regulatory arrangements. Although India does not have any separate and exclusive water law, water-related legal provisions are dispersed across various irrigation acts, central and state laws, constitutional provisions, court decisions, customary laws, and various penal and criminal procedure codes.[9] Moreover, legal provisions in the case of land, forest, and environment also have significant implications for water resources. As most of these legal provisions were enacted in the past characterized by water surplus conditions, they fail to reflect the current conditions marked by water scarcity and water conflicts. Although there were some periodic, though marginal, changes in some of these provisions, especially during the post-independence period, they are too inadequate to develop the legal system suitable for meeting the emerging challenges within water sector.

Inter-governmental Responsibility

The legal provisions related to inter-governmental responsibility in the water sector are derived from the overall constitutional division of power between the central and state governments as effected by the Indian Constitution of 1952.[10] As per Entry 17 in the State List under the Seventh Schedule of the Constitution, it is the states that have the jurisdiction over water resources within their borders. But, the powers of the states are subject to Entry 56 in the Union List that allows the cen-

[9] Even though there are well conceived proposals for the unification and simplification of irrigation acts (e.g., Jacob and Mahesh 1976) as well as for the creation of an exclusive but broad domain of water law (e.g., Singh 1991, 1992), they have not received their deserved attention from the policy-makers of India.

[10] This constitutional division of power creates three lists of sectors/activities/jurisdictions, i.e., union list where the central government has exclusive power [Article 246(1)], state list where the state governments have exclusive powers [Article 246(2)], and concurrent list where both the central and state governments exercise powers [Article 246(3)].

tral government to regulate and develop inter-state rivers and river valleys when this is declared by Parliament as a matter of public interest. The central government also has regulatory roles in the water sector vide Article 252 related to inter-state water projects as well as in terms of the Forest Conservation Act of 1980 requiring the states to get central clearance for executing ecologically sensitive water projects.

More importantly, the central government also has an important role in resolving inter-state water disputes as per the provisions under Article 262. It is in pursuance of this Article that the Parliament has enacted the Inter-state Water Disputes Act of 1956 and it is under this Act that a number of tribunals were set up to resolve water disputes among the states.[11] Besides, the Centre can also acquire legislative powers on water when two or more states desiring to have uniform water legislation request the union government with the approval of their respective assembly (Jacob and Singh 1972). Despite these legal provisions as well as other administrative and financial leverages (that we will see later) of the central government, the final legislative powers are still with the states. While this arrangement is good to address state-specific concerns, there are also serious problems as the central government is unable to provide the kind of leadership and guidance needed for reforming the legal and institutional basis of the water sector both at the national and state levels.

Water Rights

The issue of water rights as a mechanism for allocation and accountability assumes importance with increasing scarcity and conflicts both at the macro-level of regions and sectors as well as at the micro-level of distributaries, communities, and individual users. Unfortunately, India does not have any explicit legal framework specifying water rights, even though various acts have some basis for defining some form of such rights. British legislation in India during 1859-77 recognized the customary water rights of individuals and groups. But, a radical shift occurred with the Easement Act of 1882 that made all rivers and lakes the absolute right of the state.[12] While state's absolute rights can affect the development and managerial aspects of water, from the perspective of water use and its equity ef-

[11] Since the Act has failed to specify the authority to implement the decision as well as the time limit for tribunal decision, it was amended twice: first in 1980 for authorizing the central government to establish the implementation authority and then, in 2002 to specify a six-year time limit for tribunal decision (Salman 2002; Richards and Singh 2002).

[12] This position got consolidated further with the Madhya Pradesh Irrigation Act of 1931 and also had its shadow over subsequent irrigation and water supply acts enacted even in the post-Independence period. But, a number of public interest litigations have led both the Madras High Court in 1936 and the Bombay High Court in 1979 to declare that government's sovereign rights do not amount to absolute rights (Singh 1991).

fects, it is the *de facto* control over water by actual users at the micro-level that is more important.

Individual rights to both surface water and groundwater are recognized only indirectly through land rights. Thanks to the 'dominant heritage' principle implied in the Transfer of Property Act IV of 1882 and the Land Acquisition Act of 1894, a land owner can have a right to groundwater as it is considered an easement connected to the dominant heritage, i.e., land.[13] In the case of canal water, the rights to access are limited to only those having access to land in canal command areas and these rights are only use rights and not ownership rights because irrigation acts do not allow the moving of canal water to non-canal areas. Under conditions of unequal land ownership and income pattern, the practice of linking water indirectly with land and the fact of *de facto* control by better endowed persons only accentuates rural inequality and water use inefficiency.

The Model Groundwater (Control and Regulation) Bill of 1992, which was formulated and circulated by the central government for the consideration of states, though postulates a kind of groundwater permit system, it fails to set withdrawal limits (GOI 1992a). While the Bill did induce some legal initiatives in states like Karnataka, Maharashtra, and Tamil Nadu, it has not received any serious consideration by other states, even though the Bill was circulated again among the states in 1997. In view of the absence of any significant reform initiatives, the legal aspects governing groundwater resources continue to remain largely divorced from both resource realities and economic requirements (Jain 1976). Thus, the control over groundwater at the field level is governed by a *de facto* system of rights as determined by farm size, the depth and number of wells, pumping capacity, and economic power.

Accountability Provisions and Mechanisms

The two-way accountability, i.e., the individuals' accountability to each other and to the state, and vice versa could not be operationalized until a legal rights system is defined in the first place (Singh 1992:8). When law defines individual water rights, in effect, it defines not only the legal boundaries but also the physical and economic boundaries of each individual's acts and their effects on others in the context of water use. By relating rights with duties, such boundaries could be legally handled with a reasonable quantification. Thus, individual water rights system helps to trace externalities, assign payment responsibilities, minimize interpersonal conflicts, and achieve the legally grounded notion of two-way accountability. Equally important is also the issue of accountability of executives and officials to the state and the people. As most irrigation and water-related acts in India have indemnity clauses to protect the executives against the consequences of

[13] As we will see later, this provision not only constrains groundwater markets, which have merged spontaneously in many parts of India, where water is sold apart from land but also excludes those without land to have any access to groundwater.

wrong or non-implementation of stated policies, they do not provide enough incentives for the executives to be accountable either to the government or to the resource users.

The accountability of the users is sought to be influenced by negative but indirect provisions evident in penal codes and other civil/criminal procedures (Singh 1991). While some of these provisions can be used to penalize users for acts such as non-payment of water charges or illegal water diversions, there are no corresponding provisions for penalizing officials for their failure to supply water at the right time or in needed quantity.[14] Although legal provisions are necessary to infuse accountability and responsibility, they are not sufficient as their operational effectiveness depends on the kind of accountability mechanisms postulated within water laws. The accountability mechanisms currently available are both formal such as statutory, legislative, and judiciary-based mechanisms as well as informal such as decentralized and people-oriented mechanisms (Devi 1992). Of them, while the formal mechanisms are costly in terms of both money and time, the informal mechanisms such as water user associations (WUAs) and stakeholder-based basin organizations—being more accessible and responsive—could ensure accountability and dispute-resolution with least cost.

3.5.2 Water Policy

Water policy relates to the declared statements as well as the intended approaches of the central and state governments for water resource planning, development, allocation, and management. It includes statements not only on the overall policy framework but also on specific policy issues such as project selection, water pricing and cost recovery, and user and private participation. Notably, since both the general and specific policies within the water sector are also influenced often by other sectoral policies related to agriculture, public finance, and basic needs, the former cannot be dealt with in isolation of the latter. For instance, the need for attaining food self-sufficiency and consolidating the productivity gains from Green Revolution has led to the implementation of large irrigation projects, rural electrification programmes, liberal credit policies, and heavy irrigation, power, and credit subsidies. Moreover, political considerations, macro-economic necessities, and environmental concerns including natural calamities (e.g., floods and droughts) also have a strong influence on water sector policies.

National Water Policy

Although the need for a national level policy for the water sector was felt for quite some time, the immediate factor that prompted the National Water Policy (NWP)

[14] The poor recovery of water charges and illegal uses of water observed widely in many parts of India clearly suggest that these provisions against users are ineffective because they are seldom used for various political and practical reasons.

of 1987 was the unprecedented drought of 1987. The main goals of the NWP are the promotion of conjunctive use of water from surface and sub-surface sources, supplemental irrigation, and water-conserving crop pattern and irrigation and production technologies (GOI 1987). It has called for raising the canal water rates and promoting user participation in canal management. While the diagnosis of the NWP is right, its prescriptions fail to address the serious economic and institutional vacuum within which the water sector is operating. Although the NWP has recognized the need to limit individual and collective water withdrawals, it has failed to identify the institutional mechanisms necessary for defining and enforcing such physical limits. Unfortunately, the new NWP declared in 2002 has also failed to address the economic and institutional issues as it is almost a repeat of its old version, except for the recognition of the role of private sector participation and the need for a paradigmatic shift from water development to performance improvement (GOI 2002). On similar lines, several states (e.g., Andhra Pradesh, Karnataka, Madhya Pradesh, Tamil Nadu, and Uttar Pradesh) have also come out with their own water policy statements.

Project Selection Criteria

During the pre-Independence period, since the British treated irrigation projects as purely a commercial proposition, project selection policy was based on the Internal Rate of Return (IRR) criterion.[15] After Independence, there was a shift in approach. Instead of a purely commercial approach, irrigation projects were viewed as instruments for fostering socio-economic development, especially by augmenting income, employment, and food production. Consequently, the IRR, which was initially lowered to 3.9 percent in 1949, was altogether abandoned in 1958 and in its place, a rather liberal benefit-cost ratio (BCR) was used as the project selection criterion. While the Gadgil Committee of 1964 recommended a BCR of 1.5, the Irrigation Commission of 1972 (GOI 1972) allowed a BCR of just 1 for projects in drought-prone areas. However, in 1983, following the suggestion of the Public Accounts Committee, the BCR was replaced by IRR as the project selection criterion and the accepted IRR was 7 percent for projects in drought-prone and water-scarce years and 9 percent for others. Admittedly, while the reinstatement of IRR is welcome, the minimum levels stipulated are still far lower than the prevailing interest rate of about 10 percent.

Cost Recovery Policy

As to the cost recovery policy in the context of irrigation projects, successive Finance Commissions since the Fifth one have insisted on the recovery of not only the full operation and maintenance (O&M) expenses, but also a proportion of the

[15] The actually used IRR was also based on the prevailing rate of interest in the London money market as irrigation investments were often made from borrowed funds. As a result, the IRR was also periodically revised, i.e., from 4 percent till 1919 to 5 percent during 1919-21 and to 6 percent after 1921 (Sangal 1991).

interest on irrigation investment. While the Fifth Finance Commission suggested this proportion to be 2.5 percent, the two subsequent Commissions have lowered this to just one percent. Although the Eighth and Ninth Finance Commissions were satisfied just with the recovery of only the O&M costs, the Tenth Finance Commission reverted back to the stance of the Sixth and Seventh Finance Commissions, i.e., the recovery of full O&M costs plus one percent capital costs. This kind of cost recovery policy has also been supported by the Irrigation Commission in 1972, the Jakhade Committee in 1987, and the Committee on Pricing Irrigation Water in 1992. Unfortunately, the recovery policy, despite its widespread approval, was never implemented as it involves not only an upward revision in water rates but also a radical change in the very method for determining them.

Water Pricing Policy

While poor financial performance led to the recognition of the need for revising water rates, political pressure led to a delay in implementation. The main reason why farmers resist higher water rates is the general perception of water as a public good. But, the 1972 Irrigation Commission has articulated, for the first time since Independence, the private good characteristics of canal water. The Commission has also suggested that water rates have to be revised to cover, at least, 5 percent of gross income in the case of food crops and 12 percent in the case of commercial crops (GOI 1972). Despite the recommendation of this and several subsequent commissions and committees, the water charges actually recovered from farmers continue to form only a fraction of both water productivity (i.e., the difference between the productivity levels of irrigated and rainfed lands) and actual O&M costs.[16]

While the cost recovery role of water pricing policy was emphasized by many expert groups and statutory committees, the Jakhade Committee of 1987 has underlined the resource use efficiency function of water pricing policy. The Committee has suggested that the method and level of water rates that capture and convey scarcity value of the resource can both induce efficiency and ensure full cost recovery. Although many states (e.g., Andhra Pradesh, Karnataka, Madhya Pradesh, and Tamil Nadu) have recently revised water rates up to three times, the present level and the method of fixing water rates are still unable to play these dual roles. These dual roles cannot be expected unless water pricing policy forms part of an institutional and technical arrangement needed for facilitating volumetric distribution, group-based allocation, and local management (GOI 1992b; Saleth 1996).[17]

---

[16] Recovered water charges, as a proportion of O&M costs, vary from 4.02 percent in Uttar Pradesh to 73.33 percent in Orissa, whereas the same as a proportion of water productivity vary from 0.28 percent in West Bengal to 5.19 percent in Maharashtra (GOI 1992b).

[17] Volumetric allocation is not new as it was tried twice—in 1854 and in 1917 in the Ganga Canal—during the British period. Although the idea was abandoned because of its impracti-

User Participation and Privatisation

The policy towards user participation in irrigation management has evolved since 1974 when the Command Area Development (CAD) programme was initiated to hasten the utilization of the created irrigation potential through farmers' cooperation. User participation under the CAD programme was ephemeral and ineffective due to the paternalistic and *ad hoc* attitude of the bureaucracy. But, the financial crisis and physical deterioration of the irrigation systems have forced the irrigation agencies to consider farmer groups as indispensable partners in irrigation management. As a result, farmers' role in outlet level water allocation, fee collection, and system maintenance was recognized and the policy of 'turning-over' the system below the outlet level was accepted in principle and also practiced with varying degree of success (see Brewer et al. 1999). But, the major change in user participation policy occurred with the large-scale turn-over programme of canal irrigation systems in Andhra Pradesh and Madhya Pradesh. To facilitate this, these states have also amended their irrigation acts by passing special acts. In many new and upcoming projects such as the Narmada and Sardar Sarovar projects, it is categorically specified that water will be distributed only to organized WUAs.

The attention on private sector participation in the water sector has been an outcome of factors such as declining irrigation investment, poor financial performance of water projects, and the privatization of public sector enterprises initiated since the New Economic Policy of 1991. In 1995, the Union Ministry of Water Resources has constituted a high level Committee to look into the legal, economic, and technical questions related to the privatization of public irrigation projects. This committee has submitted a report favouring a gradual, selective, and stage-wise process of privatization of the irrigation sector (GOI 1995). Similarly, the state government of Madhya Pradesh constituted a committee to look into the issue of sharing the primary benefits (water supply) and secondary benefits (power generation, tourism, aquaculture, and horticulture) between the government and the private parties involved in project construction and management. This committee has recommended that the primary benefits should be retained by the government but the secondary benefits can be given to private investors (GOI 1995:135).[18] The NWP of 2002, as approved by all states, has, in fact, encouraged the reliance on private corporate sector as a potential partner for water resources development.

---

cality under the technology of that time, it has received periodic attention also during the post-independence period, especially since the Taxation Enquiry Committee of 1953.

[18] The terms of benefits offered by Maharashtra are still better as the government has allowed also a 6 percent share in total storage in addition to the full rights on all the secondary benefits (Saleth 1997).

3.5.3. Water Administration

Water administration covers the organizational, financial, and managerial structures including the regulatory apparatus and conflict resolution mechanisms, which are directly connected to the water sector. Despite considerable variations in the name and structure of water administration across states, there are a few common features such as centralized and bureaucratic character, dispersed organizational responsibilities, and weak functional linkages. Some of these shared traits become apparent as we review the overall structure of water administration.

Organizational Framework

The general organizational framework of Indian water sector can be briefly described by highlighting the key actors playing different roles both at the centre and in the states. The Union Ministry of Water Resources (MOWR), which itself evolved from the erstwhile Department of Irrigation under the Union Ministry of Agriculture, is the national organization that is responsible for the overall planning and management of the water resources in the country. The Central Water Commission (CWC), the Central Groundwater Board, and the National Water Development Agency—all under the MOWR—provide overall technical support. Research and training supports are provided by organizations such as Water and Land Management Institutes (name differs in some states), agricultural universities, and other research institutions. The Planning Commission at the national level provides the project clearance and approves the financial allocation to various water projects in different states. Other central agencies influencing the water sector, in one way or the other, include the Ministries of Economic Affairs, Agriculture and Rural Development, Environment and Forests, Housing and Urban Development, and Health.

As noted already, the actual legislative and managerial responsibilities are with the public works, irrigation, or water resource departments at the state level. Usually, there are different departments to handle the groundwater and surface water sectors in the case of most states. The main department handling the water sector also has its research and training facilities in the case of some states. There are also important organizational arrangements to achieve inter-state and centre-state coordination. These include not only various river boards charged with the responsibility of coordinating water allocation among the states in the context of a few important inter-state rivers[19] but also the National Water Resources Council (NWRC) set up in 1983 and the National Water Board (NWB) set up in 1990. The NWRC is an important policy organ in the Indian water sector as it is the apex body chaired by the Prime Minister and includes the Union Minister of Water Resources and the chief ministers/lieutenant governors of all states and union territo-

[19] Since these river boards, which were created under the River Boards Act of 1956, are bureaucratic arrangements, they are not to be confused with the stakeholder-based river basin organizations that are being advocated in recent years.

ries. The NWB—considered as the executive arm of NWRC—is chaired by the secretary of the MOWR and includes the chief secretaries of all the states/union territories, secretaries of the concerned union ministries as well as the Chairman of CWC. Local governments such as municipalities and *panchayat* unions also play an important role in drinking water supply as do the user and stakeholder groups in the irrigation sector. Pollution control boards operating usually under the ministry of environment and forests both at the centre and in the states have the responsibility for water quality aspects.

Financing and Management

Since water is a state subject, it is the states that are responsible for the financing, cost recovery, and management of all irrigation and water supply-related activities within their territory. They finance water development schemes from their own revenue, their share from the centrally collected revenue proceeds, and borrowings from financial institutions both within and outside the country. In recent years, some states such as Andhra Pradesh, Gujarat, Karnataka, Madhya Pradesh, Maharashtra, and Uttar Pradesh are also trying to mobilize funds from the private corporate sector as well as from the general public through deep-discount water bonds. As the cost recovery from irrigation and water supply projects is low and declining, many states are likely to rely more and more on private and public sources of funds. Even though the states have the major responsibility for water sector financing, the central government also plays a significant role by providing finance to states through central assistance, undertaking the construction of projects of national importance, and implementing centrally sponsored schemes such as the CAD programme in canal regions. Besides, the central government also facilitates, approves, and allocates external loans and aids to irrigation and other water supply schemes through concerned ministries including the Ministry of Economic Affairs and the Planning Commission.

Although the central government is responsible for the overall planning and coordination, it is the states that are responsible for the actual management of the water sector. The state water administration—known variously as the Irrigation Department, the Public Works Department, and the Water Resources Department in different states—is responsible for the construction, maintenance, and management of water projects. Regarding water pricing and cost recovery, the administrative systems differ in different states partly due to historical reasons.[20] Water administration in many states also has diffused or unclear administrative and functional responsibilities inapt for developing an integrated approach to water

[20] For instance, in Haryana, Punjab, Uttar Pradesh, and West Bengal, the water rate assessment is done by the water-related department, whereas the collection is done by the revenue department. But, in Andhra Pradesh, Karnataka, Kerala, Orissa, and Tamil Nadu, both functions are performed by the revenue department. This administrative problem is due to the fact that in the case of most old irrigation projects of these states, there are no separate water charges, but only a higher land revenue assessment for irrigated lands.

management. Since water management responsibilities are often with ministries dealing with public works, internal transport, or public health, they are linked with activities such as road construction and port management. But, the activities that are to be aligned such as surface water and groundwater management and irrigation and water supply provision are dispersed across departments or ministries. This problem is addressed only partially by the recent organizational reforms in states such as Andhra Pradesh, Haryana, Tamil Nadu, Orissa, Madhya Pradesh, and Uttar Pradesh. Despite the organizational reforms to the contrary, the spatial structure of water administration in most states is still based on administrative boundaries and projects rather than on well-defined hydro-geological boundaries.

Regulatory Mechanisms

While India has a relatively sound technical information base and expertise in water-related aspects, their utility at the practical level of regulation is extremely limited due to the lack of organizational arrangements for enforcement and monitoring. The top-down approach inevitable in any centralized administrative set up and the attendant inability to tap locally available informal institutional potential (e.g., water-related local customs, water sharing conventions, and monitoring and enforcement mechanisms) constrain effective enforcement of even well-conceived policies. While well-spacing norm prohibits new wells within a radius of 200 m in most parts of India, the norm can be as high as 680 m in areas with deep tubewells and serious depletion (Shah 1993:11). Similarly, there are also depth restrictions, especially for deep tubewells. For instance, in Gujarat, as per the Bombay Irrigation (Gujarat Amendment) Act of 1976 (1979), tubewell depth was limited to 45 m, particularly for the Mehsana region. Later, the limit was not only raised to 100 m but also made applicable to most parts of Gujarat. Since these spacing and depth restrictions take effect only when a farmer applies for concessional loan/well permit/electric connection, they restrict mostly the poor farmers (Dhawan 1990).

While a restricted power supply policy provides some regulatory respite, it is of little consequence in the face of large pumps and multiple wells. The effectiveness of regulations based on power tariff and supply policies is severely undermined not only by the availability of the diesel pumpset option but also by the presence of a 'kink' in the farmers' power demand.[21] Although groundwater markets are found to improve efficiency and equity in water use (Shah 1993), they could, nevertheless, accentuate aquifer depletion under current legal and institutional regimes without water rights (i.e., legalized 'water quotas') (see Saleth 1996). Thus, current legal and regulatory policies as well as these markets reinforce rather than regulate the *de facto* control of groundwater by resource-rich and influential farm-

[21] The kink in power demand emerges from the fact that the gap between energy cost and the net value of output per unit of power is very high. As long as this gap is substantial and can also be manipulated by crop choice, farmers will not reduce power consumption and hence, their power demand will be insensitive to power tariff changes (Saleth 1997).

ers. In the case of surface water resources, neither the usual policies based on water charge/supply manipulation nor the new ones based on rotational water supply system are likely to generate sufficient impact effective enough to enforce discipline in canal water use. WUAs can certainly enhance cost recovery and improve system maintenance. However, they cannot generate incentives powerful enough to enhance water use efficiency unless they operate within the framework of a legally established but locally managed system of group and individual-specific water quotas.

Conflict Resolution Mechanisms

Various arrangements exist for resolving conflicts at different levels. Water use prioritization specified in the NWP and implied in the Constitution[22] can provide a general framework for resolving inter-sectoral water allocation conflicts. But, for a more effective solution, quantification of entitlements is to accompany prioritization and both should be defined within appropriate hydro-geological and organizational contexts. Unfortunately, the issue of quantification of entitlements is often left to administrative or political decisions. The most preferred arrangement requires both the physical context of river basins and the organizational framework involving networks of stakeholders of those basins. In the case of inter-state (or inter-regional) water conflicts, the arrangement relied on often in the past involves negotiated agreements for developing and sharing water among the concerned states and regions.[23] But, when there is difficulty in reaching a negotiated settlement, then, the concerned parties can rely on the tribunal established by the central government under the provisions of the Inter-state Water Disputes Act of 1956.[24]

As the pressure for the renegotiation of existing agreements builds up with increasing water scarcity, the role of the tribunal mechanism is likely to increase in the future. Although the tribunal awards settle the dispute by quantifying the water claims, they involve a lengthy process to reach a final settlement. Even the six-year time limit specified by the 2002 amendment is too long. Besides, since the

---

[22] For instance, the constitutional provisions relating to fundamental rights to life (and clean environment) are often used as a basis for assigning top priority for drinking and domestic uses as well as irrigation and ecological water needs.

[23] There are about 58 independent water-related agreements among states concluded in the past—39 related to joint projects and 19 related to sharing of river waters—and all of these are under heavy pressure for renegotiation due to the increasing water requirements of concerned parties.

[24] Under this Act, the central government has so far set up five tribunals and three of them have come out with amicable decisions (Krishna in 1976, Godavari in 1979, and Narmada in 1979). These include also the tribunal dealing with the politically most sensitive Cauvery River dispute where only an interim award was given and even that is strongly contested by one of the concerned states.

implementation of tribunal awards can be contested in the Supreme Court, it is crucial to provide legal binding to final awards. Although market or negotiation-based arrangements are not tried much in India, there is considerable potential for them. States can be encouraged to purchase and sell water either on a payment basis or on a barter basis (i.e., exchange of water for power or foodgrains). There are also cases such as the Krishna water transfer for Chennai (Madras) city where Tamil Nadu has paid for the entire project costs (implicitly paying for water).

As to the mechanisms for resolving water-related conflicts at the micro-level, there are a few traditional and informal village level institutions (e.g., tank-level organizations in Tamil Nadu). By rejuvenating these informal institutions as well as by strengthening the formal arrangements such as the *panchayat* institutions and WUAs, it is possible to build a more effective and accessible conflict resolution mechanisms at the grassroots level. However, the middle level conflicts across communities within a river basin or canal system (e.g., upstream users vs. downstream users) as well as the conflicts between irrigation and water supply agencies are still rampant for want of proper forums for resolving differences. The arrangements forming part of the bureaucracy (e.g., divisional engineer or district collector) not only remain inaccessible for all but also turn out to be artificial as free expression of mutual concerns becomes difficult. Basin or system level stakeholders' association involving user groups and officials can be a very useful forum for promoting both conflict resolution and accountability.

3.6 Water Institutional Structure: Micro Perspective

Although colonial policies as well as the post-colonial expansion of government bureaucracy have severely eroded most of the indigenous water institutions, India still presents a rich variety of locally managed water-related institutions, especially in the water-scarce areas of Bihar, Maharashtra, Rajasthan, and Himachal Pradesh (Datye and Patil 1987; Sengupta 1993). These institutions are in the form of informal customs and conventions for water sharing as well as community-based organizations for water management. Although these institutions remain largely independent of formal water institutions and operate only at the periphery of the formal water sector, they can still provide very valuable insights for designing the kind of institutional mechanisms that are needed for filling the organizational vacuum existing at lower echelons of water management. In addition to these traditional arrangements, new forms of informal arrangements for sharing water and irrigation service have also emerged in recent years, especially in the groundwater regions. They are the rental markets for irrigation assets, groundwater markets, and myriad forms of water contracts. As these informal institutions—both traditional and emerging ones—have significant implications for the operation and performance of formal institutions, they deserve our attention.

3.6.1 Localized Institutions

Even though the legal system in India has not formally specified any water rights system, there are evidences for the operation of rudimentary water rights systems capable of being developed into an effective water rights system. Informal water rights—both individual and group—have existed in India since ancient times (Siddiqui 1992) and continue even today, albeit in a much weaker form, in many tank irrigation systems of South India (Vani 1992). A few South Indian irrigation systems have informal and prioritised water rights not for individuals, but for different distributaries or command segments (Vaidyanathan 1985:63-64). The 200-year old *Phad* system operating in the Panijhra River area of the Dhule district and the the *Pani Panchayat* system being practised in parts of Purandhar taluk of Pune district, Maharashtra (Datye and Patil 1987:42-123) have the potential for creating a user-managed water rights system.[25]

In the deltaic regions of Orissa and West Bengal as well as in parts of Bihar and Madhya Pradesh, there exists an officially-granted non-transferable long-term water lease system designed for encouraging farmers to use surface water (GOI 1976:65). More important and interesting is the *Shejpali* (water distribution roster) system being practised in the canal commands of western Maharashtra. Under this system, the canal authorities issue 'water passes' on the basis of an application from farmers in the command on a 'first-come, first-serve' basis. The duration of these water passes varies from six years to a single crop season and their priority varies somewhat directly with their duration. That is, longer duration passes have seniority over shorter duration passed (Gandhi 1981; Rath and Mitra 1989). But for their non-transferability and quantitative specification, these water passes system resembles very closely the water permit system of the mid-western states of the US. Besides these relatively well-known systems, there are also a variety of informal organizational arrangements and water allocation conventions that can be observed both in the mountain and foothill regions and in the tank irrigated areas.

3.6.2 Rental Markets for Irrigation Assets

Increasing value of irrigation water as well as land and resource-related bottlenecks for private investment in irrigation assets have led to the emergence of the phenomenon of rental markets for irrigation assets such as wells and pumpsets.[26]

[25] Notably, under the *Pani Panchayat* system, the water share of users is based not on farm size but on water needs. Since water needs are calculated at about half an acre worth of irrigation water per person, the allocation criterion is divorced from land ownership, but linked closely with family size (Singh 1991:35; Vani 1992:9-10).

[26] This phenomenon involves the use of dugwells or tubewells and hence, the transfer of water in the case of lift irrigation in groundwater regions. But, in surface water-based lift

Since these markets allow farmers to irrigate their farms by renting the irrigation assets from their neighbours, they contribute both to equity in water use and better utilization of irrigation assets. According to national sample survey information [National Sample Survey Organization (NSSO 1984, 1985), about 10 percent of the total pumpsets in the country are involved in pumpset rentals. Since about 63 percent of these rentals occur with dugwells/tubewells with electrically powered and permanently fitted pumps, it seems that the majority of the rentals involve water transfer as well. This is particularly so for the Indo-Gangetic and hardrock states. Since the rest of the rentals occur in the case of other water sources where pumps can be physically moved with little cost, it seems they occur independently of water transfers. This is particularly so in the deltaic states such as Bihar, Orissa, and West Bengal (Saleth and Thangaraj 1993).

The expanding phenomenon of pumpset rentals is an indication of the existence of surplus pumping capacity, particularly in the case of diesel pumpsets. In condition where irrigation assets account for 16 percent of rural assets and up to 40 percent of private fixed capital formation in agriculture (Saleth and Thangaraj 1993), the underutilization of irrigation assets is a serious issue. The informal institution of rental markets has emerged essentially as a response of the farmers to address this problem. This phenomenon also brings forth certain conceptual and institutional implications. For instance, when there is the physical movement of pumpsets from one water source or location to another, they are just pure pumpset rentals as there is no water transfer between farms. [27] But, in the hardrock and Indo-Gangetic regions where pumpsets—mostly electricity-based—are permanently installed with dugwells/tubewells, pumpset rentals necessarily involve water transfers and hence, the rental activity is institutionally linked with groundwater markets (Saleth 1996).

3.6.3 Groundwater Markets

Despite their localized, fragmented, and uneven nature across regions, groundwater markets are growing in magnitude and gaining in significance. While water selling practices in India are traced to the 1920s, more systematic documentation of this phenomenon started only since the late 1960s. Their characteristic features are that they occur without any formal water rights system and involve no sacri-

irrigation systems where pumpsets can be physically moved from one farm to others with little cost, rentals can occur independent of the use of wells and hence, water transfers.

[27] In their survey of few West Bengal villages, Kolavalli and Atheeq (1990:26) report such pure pumpset rentals where diesel pumpsets are leased on a seasonal basis with the lease rate per crop season varying between Rs. 1,200 and Rs. 1,500.

fice of self-irrigation.[28] While there is no systematic national level study on the magnitude of water selling, based on his studies in Gujarat and Uttar Pradesh, Shah (1993:250) has projected the area irrigated through groundwater markets to be up to 50 percent of the total gross irrigated area under private lift irrigation. Understandably, there are considerable variations across regions. While the area irrigated through groundwater markets is projected to be up to 80 percent for north Gujarat (Shah 1993:205), the corresponding figure for Uttar Pradesh is about 60 percent (Shankar 1992:12). But, in Vaigai basin, Tamil Nadu, the area under purchased water was reckoned at no more than 30 percent of the total irrigated area (Janakarajan 1993). In contrast, there are also studies which report no water selling at all in their respective sample areas (see Shah 1993:55).

Coming to the dominant technical and institutional features of water markets, although their geographic locus is limited by the physical characteristics of the groundwater and farming systems, their size is often enlarged by elaborate underground pipeline networks. But, such market expansion is confined to very few regions and often leads to an unbalanced market structure as it adds more buyers than sellers to the market. Since buyers are mostly small farmers, they have a weak bargaining position. For instance, a typical water sale-purchase matrix in Uttar Pradesh reveals that the demand side is dominated by smaller farms with less than 2 acre as they account for 81 percent of the total area under purchased water (Shankar 1992:33). In view of the monopolistic or oligopolistic tendencies in these markets, not only are water rates several times higher than the pumping cost, but also price and non-price discriminations remain pervasive. The root cause for the sub-optimality of these groundwater markets lies not so much in their economic and organizational aspects but in the legal and institutional vacuum within which they operate at present. As a legally instituted and locally managed water quota system defined within an ecologically consistent overall withdrawal limit can provide more powerful incentives for water use efficiency, it could eliminate their negative effects, but magnify their positive benefits (Saleth 1996).

3.6.4 Water-Based Contracts and Conventions

There are a variety of water-based tenancy contracts. Although these contracts are often treated as part of groundwater markets, they need to be differentiated as they involve the use of other resources such as land, labour, capital, and other farm inputs that are governed by distinct sets other institutions. For instance, there are two distinct types of such contracts in Kheda district, Gujarat. They are: (a) a two-party contract where water sellers provide irrigation, share 50 percent of cash expenses (except labour costs), and claim 50 percent of output and (b) a three-party contract where water seller, land owner, and labourer share equally the cash ex-

[28] Under these conditions, there is not only a conceptual issue (i.e., whether the sellers are selling water or excess pumping capacity) but also an economic question (i.e., the opportunity costs are undefined when there is not sacrifice of self-irrigation).

penses as well as crop output (Shah 1993:51-52). Similarly, in Karimnagar district, Andhra Pradesh, water sales occur within labour contracts, crop sharing contracts, and crop and input sharing contracts. Notice that these contracts represent not only an institutional evolution of crop sharing within the context of WMs but also link WMs with other rural input/output markets.

Apart from these water-based tenancy contracts, the pricing methods are also accompanied by certain informal conventions and contractual obligations with considerable implications for water use efficiency and risk-sharing (see Kolavalli and Atheeq 1990:38-40). For instance, the area-based method involving crop shares, though provides lesser incentive for water conservation than the method involving hourly rates, allows risk-sharing between buyers and sellers and also involves some informal contractual obligation for sellers to provide irrigation for the whole season. Besides, in the case of both the area and per irrigation rates, there are also mutually agreed upon conventions (e.g., the level or intensity of irrigation constituting 'full irrigation') to avoid conflicts and water overuse. Since water sharing contracts also lead to input/output selling contracts, there are also evidences for inter-linked input and output markets in rural areas (Janakarajan 1993).

3.7 Evaluating the Performance of Water Institutions

The overall performance of water institutions depends not only on the individual effectiveness of the legal, policy, and organizational components and their underlying institutional aspects but also on the joint performance as determined by the strength of the structural and functional linkages among these institutional components and aspects. Besides these internal and structural features, the performance of water institutions depends also on the general institutional environment within which they evolve and operate. While this perspective of evaluating water institutional performance is consistent with our institutional decomposition-based analytical framework, its empirical translation with objective information presents a major challenge. Nevertheless, this approach is amenable for empirical application when it is possible to use perception-based judgmental information collected from a carefully selected sample representing a cross-section of water sector stakeholders (Saleth and Dinar 2003).

When perception-based data cannot be collected due to the lack of time and resources, another alternative could be the reliance on the learned and balanced judgements of the researchers themselves. The key aspect for consideration in this respect can be the overall progressiveness of water institutions as captured in terms of their adaptive capacity, amenability for innovation, openness for change, and ability to tackle emerging and future water sector challenges (Saleth and Dinar 1999). As we reflect these criteria over our brief description of the structure of Indian water institutions both at the micro and macro levels, we can derive few general and qualitative results. For instance, we see the micro-level institutions (e.g., groundwater markets) are relatively more responsive to changing local needs whereas macro-level institutions continue to lack the required extent of flexibility as changes are resisted by factors ranging from path dependency to political economy obstacles. Even within formal institutional components, water policy is relatively more responsive as compared to water law and water administration. The policy changes (e.g., water policy statements by the national and a few state governments) are mostly in the nature of a political expression of intentions and aspirations rather than leading to practical attempts for the implementation of the stated policies. As a result, they are politically easy to accomplish, but, operationally ineffective in ensuring the intended socio-economic benefits. However, policies related to more substantive aspects such as water pricing and inter-regional water transfers have not changed much. Similarly, in the sphere of water administration, the changes such as the administrative reorganizations are mostly superficial whereas substantive aspects such as the oversized and functionally unbalanced staffing pattern, creation of basin organizations, and making water-related departments financially autonomous have not changed to the desired extent.

From another perspective, water sector performance can also be used as a proxy for understanding water institutional performance. While the performance criteria applicable at the project and system levels are relatively more rigorous (e.g., Bos 1997; Burt and Styles 1997), those needed for evaluating the performance of water sector as a whole are more indicative rather than conclusive.[29] Recognizing this fact, we follow a simple yet meaningful approach of evaluating the overall performance of Indian water sector in terms of three gaps: the physical gap, financial gap, and economic/incentive gap (Saleth 1996). The physical gap can be evaluated in terms of both the gap between water resource potential and its utilization as well as the gap between water demand and supply. The financial gap can be indicated in terms of the gap between water sector investment and cost recovery. Similarly, the economic/incentive gap can be reckoned in terms of the gap between the average value of water and the water rate being charged. Finally, we note that these three gaps are evaluated essentially in the particular case of irrigation sub-sector that uses up to 80 percent of total water withdrawals.

The utilization gap is already indicated in an earlier section as part of our discussion on water institutional environment. The demand-supply gap that which has already assumed serious proportions at local and regional contexts (e.g., Cauvery Basin, in southern India) is also growing at the aggregate level. In the particular context of irrigation sub-sector, the demand gap is very serious as the ac-

[29] Even with well-developed objective performance criteria, water sector performance cannot be evaluated in all its dimensions due to the presence of crucial subjective aspects of performance such as the smoothness of water transfers and the adaptive ability of water institution. While proxies like the number of water sector conflicts can be used to capture this aspect, it is not clear, for instance, how one can factor into the evaluation the relative seriousness of such conflicts. Despite these problems, we can evaluate.

tual irrigation potential created so far is only about 88 million hectare (mha) as against the gross sown area of 185 mha.[30] The problem is going to be complicated further by the declining share of irrigation caused by the increasing demand pressures from other water sub-sectors. From the perspective of institutional performance what all these forms of physical gaps suggest is the fact that current institutions, particularly those related to inter-basin transfers, inter-sectoral allocations, and conflict resolution, are weak requiring urgent reforms.

The financial gap in the water sector can be approximated by the difference between the total investment costs and total revenue in the canal irrigation sector. The total investment in canal irrigation during 1951-2000 is estimated to be Rs. 790 billion at current prices (GOI 2000). Even if we assume a simple rate of 8 percent to account for both interest and depreciation, the annual financial cost of canal irrigation provision comes to about Rs. 63 billion. Although we do not have published information on the revenue from canal sector, going by the estimate of the Committee on Pricing Irrigation water (GOI 1992b), we can reckon that the total revenue at present can be in the range of about Rs. 3-5 billion. Such a magnitude of financial gap clearly shows that the institutional aspects such as cost recovery and water pricing policies as well as the organizational mechanisms involved in water charge collection are performing rather poorly.

Since the economic/incentive gap shows the extent that water charges remain below the economic value of water, it is clearly linked with the financial gap, especially water pricing and cost recovery policies. Water rates are not only very low but also rarely revised. As a result, they are related neither to productivity nor to provision cost. They cover hardly 5 percent of water productivity (i.e., the difference between the average productivity levels of irrigated and rainfed lands) and 8 percent of O&M costs. Paradoxically, since even the low water charges are not fully recovered, the arrears are also accumulating over time in most states. Besides these effects on the financial side, the low and uneconomic water rates also lead to an incentive problem causing widespread water use inefficiency. The incentive gap can be approximated by the gap among water productivity, supply cost, and water rates. In the context of canal regions, while water productivity is reckoned in the range of Rs. 714-5812/ha and supply cost is estimated to be in the range of Rs. 90-603/ha, water rates were in the range of Rs. 6-1,000/ha) (GOI 1992b). The incentive gap indicates not just the poor performance of pricing and cost recovery policies but also the absence of institutional conditions necessary for volumetric allocation such as the water rights including their legal and organizational requirements.

[30] Even if the estimated ultimate irrigation potential of 139 mha is fully developed, India will still continue to have this irrigation gap as the gross sown area is expected to grow further to 210 mha by 2025.

3.8 Institutional Change: Nature, Extent, and Causes

The review of water institutional structure clearly shows that legislative powers, technical capabilities, planning skills, and operational responsibilities are dispersed across government layers. As a result, water institutions in India remain legally weak, functionally disjoint, and regionally uncoordinated. While the physical stress, financial crisis, and performance gaps have clearly exposed the legal, policy, and administrative weakness of water sector, myopic political issues and administrative resistance continue to delay the implementation of the reform blueprint developed jointly by the GOI and the World Bank (World Bank 1998). This does not mean that there are no changes in the institutional sphere both at the central and state levels.

3.8.1 Nature and Extent of Changes

At the national level, the central government has constituted a number of national committees/commissions to review specific water policy issues as well as plan for the long-term development of the water sector. Among them, the notable ones are the Committee on Pricing Irrigation Water (1992), Committee on Private Sector Participation in Major and Medium Irrigation Projects (1995), and the National Commission of Integrated Water Resources Development Plan (1997). While the first one has revived the attention on the rationalization of water rates, volumetric water allocation, and system modification, the second one has documented the rationale, feasibility, and actual state level initiatives for involving private corporate sector, especially in the construction and modernization of irrigation schemes. The National Commission has developed a national master plan for the water sector by synthesizing and updating similar plans prepared earlier by the CWC as well as investigated the economic, technical, and institutional issues in the water sector from a national perspective.

Although these three initiatives seem to be like any other routine committees/commissions, they are a very important part of the long process of water sector reforms in India. In fact, some of their suggestions are reflected in the new NWP declared in 2002 with the consensus and approval of all states. It is true that its final version is a substantially watered down version drafted in 1999 in view of the reservations of some states, especially on statements related to water sharing. But, it also signifies a consensus on two important policy issues that represent a somewhat radical departure from the past. These relate to the agreement on private sector participation in water resource development and on the need for a paradigmatic shift from water resources development to their allocation and management.

Besides the new NWP, there are also notable changes on the legal front—some initiated by the central government and others by public interest litigations by concerned individuals/groups. To avoid the usual delay in tribunal decisions, the Inter-state Water Disputes Act of 1956 was amended in 2002 for mandating a six-year limit for the tribunal to give its final award. While this amendment helps to reduce the usual time involved in tribunal decisions, the gravity of some water disputes (e.g., Cauvery water dispute) necessitates not only a much shorter time for tribunal decisions, but also a different organizational arrangement for resolving conflicts. It is for this reason that the Cauvery River Water Authority—a new kind of institutional arrangement for dealing with inter-state water disputes—has been created to deal with the water allocation conflict among the four states in the Cauvery River basin.[31] While public interest litigations have invoked the Supreme Court to order the closer/relocation of many polluting industries on the banks of Yamuna and Ganges rivers, presidential intervention has invoked the same to order the central government to investigate the feasibility of linking major rivers.[32]

At the state level, there are also noteworthy changes, especially in the case of Andhra Pradesh, Karnataka, Madhya Pradesh, Maharashtra, Tamil Nadu, and Uttar Pradesh. These changes can be seen both in the organizational spheres (e.g., administrative reorganizations including the creation of basin organizations in states such as Tamil Nadu and Uttar Pradesh) as well as in the policy spheres (e.g., declaration of water policy statements by most states). There are also more substantive changes in a few states. For instance, Andhra Pradesh and Madhya Pradesh have gone for a state-wide programme for the transfer of the management responsibilities of almost all canal irrigation below the outlet and minor levels.[33] Notably, such transfers occurred with full legal backing under specifically enacted legislations. With the transfer of irrigation management responsibilities and the organization of regular elections and training to the office-bearers of WUAs, these states could see farmers accepting an increase in water charges up to three times and undertaking cost recovery and maintenance functions. Although irrigation management transfer in other states was not of that magnitude observed in these two states, they demonstrate both policy commitments and concrete actions for the devolution of allocation, cost recovery, and maintenance functions to WUAs.

Another important change at the state levels pertain to the creation of autonomous corporations by Karnataka and Maharashtra for mobilizing public funds as well as the initiatives of Andhra Pradesh, Gujarat, Madhya Pradesh, and Ma-

[31] This entity—patterned after the Murray-Darling River Basin Organization of Australia—is unique in Indian context as this is the first time that a basin organization is chaired by the Prime Minister with the chief ministers of all the concerned states as members.

[32] This inter-basin transfer scheme—popularly described as the 'Garland Scheme'—has been proposed for a longtime not just by engineers and politicians but even by poets and other social thinkers.

[33] For instance, under its Farmers' Participation in Irrigation Management Act of 1999, Madhya Pradesh has transferred outlet management responsibilities of all surface irrigation schemes to 1,470 legally constituted and formally elected WUAs—466 in major, 158 in medium, and 846 in minor schemes.

harashtra for soliciting corporate investments in the water sector. In 1994, Karnataka has formed the Krishna Bhagya Jal Nigam Limited (KBJNL) under the Companies Act with the specific purpose for mobilizing public funds for developing the Upper Krishna Project. Thanks to the high return (about 17.5 percent) and government guarantee, the water bonds issued by the KBJNL during 1995-99 have fetched an unexpected sum of Rs. 23 billion.[34] Almost similar is also the case with the Maharashtra Krishna Valley Development Corporation (MKVDC) floated by Maharashtra in 1996. The water bonds of MKVDC also promise a 17.5 percent return payable half-yearly with a maturity period of six years for the first 50 percent of the value and seven years for the remaining 50 percent of the value. Besides, these bonds also allow an up-front discount of 2.5 percent of their face value. As a result, the first public issue of MKVDC undertaken in July 1996 has fetched an amount of Rs. 4.28 billion as against the original target of only Rs. 1.5 billion. Encouraged by the overwhelming response, the Maharashtra government has plans to mobilize over Rs. 36 billion through MKVDC (Saleth 1999).

In addition to its efforts through MKVDC, the government of Maharashtra is also trying to tap direct investment from the private corporate sector. For instance, in 1996, it has invited private bids for 52 irrigation projects worth Rs. 150 billion. On similar lines, the governments of Andhra Pradesh, Gujarat, and Madhya Pradesh have also tried to tap the private sector both for the construction and modernization of few water projects (GOI 1995). Interestingly, in its Agricultural Policy Resolution of 1995, the Government of Karnataka (GOK) has indicated its willingness to grant financial autonomy to the Irrigation Department by converting it into a corporation and making the farmers co-owners of the irrigation structure with 'water equity shares' (GOK 1995).[35] Finally, for completeness, let us also state that the changes in the informal institutions at the grassroots level are substantial, which are fomented by both macro-level changes (e.g., irrigation management transfer) as well as micro-level changes in conventions and contractual arrangements (e.g., irrigation asset and water sharing conventions).

3.8.2 Explaining Changes within a Transaction Cost Perspective

As we reflect on the observed changes within the institutional transaction cost framework outlined earlier, we can find some interesting insights into the role and relative significance of various factors motivating these changes in the Indian context. The socio-economic consequences of widespread groundwater depletion, ecological costs of large scale water development projects, storage loss from siltation, and command area loss to waterlogging and salinity have remained the fun-

[34] Since a high credit rating of these bonds has led to over-subscription, the KBJNL has revised down the return to only 14.25 percent in July 1998.

[35] The other states where this sort of more is likely to take place in the near future are: Haryana, Rajasthan, and Tamil Nadu (Gulati and Mainzen-Dick 1996).

damental factors providing a strong economic motivation for water institutional reforms. Unfortunately, the financial and performance crises of the water sector have failed to gather the political economy thrust needed for prompting concrete actions. From an institutional transaction cost perspective, what this means is the fact that although the opportunity costs of inaction were high in economic terms, the perceived political costs of taking actions were still higher to undermine the reform initiatives.

Fortunately, there were a number of developments—mostly exogenous to water sector—since mid-1970s that have not only reduced the political costs but also magnified the opportunity costs of water institutional reforms. For instance, the earlier practice of making water policy subservient to food self-sufficiency policy has lost its relevance when India eliminated food imports in 1971 and started building a comfortable buffer stock (often going beyond 20 million tons in recent years). Although the first NWP was prompted by a water-related phenomenon of the widespread drought of 1987, it is the political and media fallout associated with this natural event that, in fact, galvanized the necessary political will to declare even such a simple non-binding policy statement. But, the transaction cost calculus with respect to water institutional reform changed rather dramatically with the macro economic crisis of the late 1980s and subsequent declaration of the New Economic Policy of 1991 focusing on financial discipline, economic liberalization, and liquidation of public sector enterprises.

The major impact of the New Economic Policy on water sector has been a radical decline in its budgetary share. For instance, the share of irrigation sector alone has declined from 23 percent of total plan expenditure in the 1950s to 7 percent in recent years. In the wake such budgetary cuts, water-related departments were forced to take a harder look into the ways of cost saving and fund raising from the sector itself. At the same time, farm lobbies that were resisting water rate revision have also realized that farm income is getting increasingly affected by unreliable water supply from poorly maintained irrigation systems. It is these macro economic conditions and their micro economic consequences that have magnified the opportunity costs of reforms from the perspective of both the government and water users. The situation, therefore, has become favourable to revise water rates, involve farmers in cost recovery and system maintenance, and consider broader water sector reforms. It is in this particular environment characterized by economy-wide reforms that the central government constituted the committees for look into issues such as water pricing and private sector participation and the state governments went for the creation of autonomous corporations and the promotion of private sector participation in water sector.[36]

---

[36] These changes in the case of Andhra Pradesh, Karnataka, and Maharashtra also have an ulterior motive as they use these unconventional means of financing water projects so as to establish their control over water resources in the Krishna basin before the tribunal award comes for renegotiation.

The revival of interest on many policies that were considered once as anathema or impractical (e.g., irrigation privatization, volumetric water allocation, water rights, and moving water into the concurrent list) indicates the changing balance in the policy debate. This is certainly a positive development and augurs well for the prospects of more substantive reforms in future. From another perspective, the economic and trade liberalization policies initiated since 1991 have also produced significant scale economies in terms of their synergetic effects on water sector reforms. Since water sector reforms form part of an economy-wide reform, the political economy costs of the former became a small proportion of the latter.[37] Meanwhile, international lending agencies (e.g., World Bank and Asian Development Bank) and research and technical organizations (e.g., International Water Management Institute (IWMI) and Food and Agriculture Organization) also have considerable influence on the nature and direction of water sector reforms. For instance, most of the organizational reforms, including the promotion of basin-based organizations observed in states such as Andhra Pradesh, Tamil Nadu, Orissa, and Uttar Pradesh, were introduced under different World Bank-funded projects. Similarly, international best practices and research-based knowledge produced by organizations such as IWMI have also played a catalytic role in reducing the transaction costs of irrigation management transfer programmes, especially in Andhra Pradesh and Madhya Pradesh.

While Indian water sector is gradually coming out of the bureaucratic grip and myopic political considerations, it has not yet fully matured to be influenced mainly by economic and technical forces. Unfortunately, the divisive role of political factors is likely to increase, especially on the issue of inter-state water sharing, as most rivers in India are shared by two or more states. As the basin resources are fully appropriated, additional claims will be politically more acrimonious unless institutional arrangements are created to catalyse negotiated settlements and mutual agreements. But, the issue has become complicated by the proposed 'Garland Scheme' for linking rivers, especially when the central government lacks the legal powers and political will to implement the scheme, though has the technical and financial arrangements to complete most of the feasibility studies for various links. It is in this political vacuum and indecisive environment that the Supreme Court of India has been invoked by several public interest litigations to require the government to report on the feasibility and prospects of this Scheme. This legal injunction by the Supreme Court has significantly reduced the political transaction costs for the Scheme. Whatever be the logic and arguments of both the proponents and opponents of the Scheme, the fact still remains that it is one of the long-term options for India to address its future water shortage problems.

[37] This fact clearly underlines the transaction cost implications and strategic significance of timing and packaging the water sector reform so as to make it coincide with and form part of a larger economy-wide reform programme.

3.9 Concluding Remarks

Despite our selective and eclectic review of Indian water institutional structure and the indirect and tentative way of evaluating its performance, our attempt does leave few interesting observations with considerable implications for both institutional economics theory and water sector reform strategies. The present structure of water institutions in India, as reviewed in terms of some of their major institutional components and aspects shows that it is far from the system that is required for the country to meet its water challenges at present and in future. This observation is reinforced further by a below standard performance of water institutions, as evaluated indirectly in terms of the overall physical, financial, and economic performance of water sector. However, there are notable reform initiatives to strengthen the water institutional structure and improve its performance. These initiatives, undertaken especially since 1991, provide observational evidence for the fact that reform benefits (or, opportunity costs of inaction) are exceeding the corresponding economic and political transaction costs. But, the fact that institutional changes are uniform neither across institutional components nor across water sub-sectors suggests that both the opportunity and transaction costs vary considerably by institutional and sectoral contexts.

The nature, extent, and coverage of institutional reform clearly provide evidence for the powerful effects that exogenous factors (e.g., economic liberalization policies, political forces, international financial and research institutions, and natural calamities) have on the opportunity and transaction costs of institutional change within water sector. Notably, the initiatives undertaken initially involved only the transaction cost-wise easier and ceremonial options (e.g., declaration of water policy, constituting committees, and marginal legal amendments). However, those undertaken in recent years involved politically difficult and substantive options (e.g., administrative reforms, basin organizations, irrigation management transfer, and reliance on autonomous corporations and the private sector). But, India is yet to move to the stage of embarking on real reforms (e.g., review of centre-state relation in water sector, declaration of an exclusive water law, creation of water rights system at various levels, and administrative reforms for water subsectoral coordination, staff resizing, and balanced functional specialization). Understandably, these reform options involve heavy economic and political transactions costs. Although these costs are lower than the potential performance benefits, the differential weights assigned by political leaders often distort the transaction cost calculus.

While India has to go a long way to set right its water institutional structure, from the perspective of a stage-based process of institutional change, the changes observed so far do signify that India is on the threshold of ushering in the substantive phase of institutional reform.[38] This observation is based on four reasons.

[38] As per the theory of institutional change proposed by Saleth and Dinar (2003), the change process involves circularity with four distinct stages, each with different dimen-

First, although the observed changes are slow, partial, and inadequate, their direction and thrust are on desired lines. Second, the nature and tenor of these changes –both already observed and those proposed in an approved reform blueprint (World Bank 1998)—indicate a clear commitment of the central and state governments to move ahead with the process of institutional change. This commitment is likely to be strengthened further by additional pressures from factors both endogenous and exogenous to the water sector. Third, as the already initiated reforms begin to yield benefits, strengthen pro-reform constituencies, and reduce the technical and political costs of transacting additional reforms, the incentive balance within the institutional transaction cost framework is likely to move toward further reforms. Finally, but, more importantly, since the path dependency properties of institutional change will ensure that it is costlier to return rather than to go ahead in the reform path, further reforms are more likely to be undertaken. As a result, the reform process, though can be delayed, can neither be stopped nor reversed.

While the prospects for undertaking higher level reforms are brighter for India, there is an indispensable need for a clear and long-term strategy for reform implementation. In this respect, some of the key insights from cross-country analyses of water institutional reforms (e.g., Saleth and Dinar 2000, 2003) can be used both for reform design and implementation. These involve the issues of timing to strategically exploit the synergy effects from exogenous factors, scale-related effects of institutional inter-linkages (e.g., links between WUAs and pricing policy and WUAs and volumetric allocation), and institutional sequencing and packaging (e.g., undertaking easier reforms first and implementing related programmes together). As the transaction cost declines and political balance improves as we move on the institutional change continuum, it is prudent to pursue a logically linked sequential reform strategy wherein water sub-sectors and institutional components are prioritized in terms of their performance impact, facilitative roles for downstream reforms, and political acceptability. Since such a strategy can exploit better the synergies from both institutional inter-linkages and exogenous factors with proper timing, packaging, and sequencing of water sector reforms, it has a better chance of success.

3.10 References

Bos MG (1997) Performance indicators for irrigation and drainage. Irrigation and Drainage Systems 11(2):119-137

sions of time and configuration of drivers. These stages are marked by mind change in terms of perceptional convergence among stakeholders as induced by both objective and subjective factors, procedural and ceremonial changes, actual and substantive changes, and the realization of actual performance benefits.

Brewer J, Kolavalli S, Kalro AH, Naik G, Ramanarayan S, Raju KV, Sakthivadivel R (1999) Irrigation management transfer in India: policies, processes, and performance. Oxford & IBH Publishing Co Pvt Ltd, Oxford

Burt CM, Styles S (1997) Irrigation modernization study, appendix: irrigation indicators. World Bank-IPTRID-IIMI, Washington DC

Coase RH (1937) The nature of the firm. Economica 4(2):1-44

Datye KR, Patil RK (1987) Farmer managed irrigation systems: Indian experience. Centre for Applied System Analysis in Development, Bombay

Devi PR (1992) Accountability in water resources management. In: Singh C (ed) Water Law in India. Indian Law Institute, New Delhi

Dhawan BD (1990) Studies in minor irrigation. Commonwealth Publishers, New Delhi

Gandhi PR (1981) History and practice of management of irrigation waters in Maharashtra. Water and Land Management Institute, Aurangabad

GOI (1995) Report of the Committee on Private Sector Participation in major and medium irrigation projects. Ministry of Water Resources, New Delhi

GOI (1992a) Model bill to regulate and control the development of groundwater. Ministry of Water Resources, New Delhi

GOI (1992b) Report of the Committee on Pricing Irrigation Water. Planning Commission, New Delhi

GOI (1976) Report of the National Commission on Agriculture, Part V: Resource Development. Ministry of Agriculture, New Delhi

GOI (1972) Report of the Irrigation Commission, New Delhi

GOK (1995) Draft Agricultural Policy Resolution. State Planning Board, Bangalore

Gulati A, Mainzen-Dick R (1996) Reforms options in Indian canal irrigation, paper presented in the Workshop on Institutional Reforms in Indian irrigation. National Council of Applied Economic Research, New Delhi, November 6

Iyer RR (1994) Indian federalism and water resources. Water Resources Development 10(2):191-202

Jacob A, Mahesh G (1976) The unification and simplification of irrigation laws in the country. Bhagirath, 23(1):102-107

Jacob A, Singh SN (1972) Law relating to irrigation. In: NM Tripathi Private Ltd, Bombay

Jain SC (1976) Legal aspects of groundwater management. Journal of the Indian Law Institute 23(1): 181-189

Janakarajan S (1993) Economic and social implications of groundwater irrigation: some evidence from South India. Indian Journal of Agricultural Economics 48(1): 65-75

Kolavalli S, Atheeq LK (1990) Groundwater utilization in two villages in West Bengal. Centre for Management in Agriculture, Indian Institute of Management, Ahmedabad

North DC (1990a) Institutions, institutional change, and economic performance. Cambridge University Press, Cambridge MA

North DC (1990b) A transaction cost theory of politics. Journal of Theoretical Politics, 2(4): 355-367

NSSO (1984, 1985) Sarvekshana 8(2):S1-S104 and 9(1):S1-S107

Planning Commission (1993) Report of the Expert Committee on Estimation of Proportion and Number of Poor, New Delhi

Rath N and Mitra AK (1989) Economics of irrigation in water-scarce regions. Artha Vignana 31(1):1-129

Richards A, Singh N (2002) Inter-state water disputes in India: institutions and policies. International Journal of Water Resources Development 18(4):611-625

Saleth RM (1996) Water institutions in India: economics, law and policy. Commonwealth Publishers, New Delhi
Saleth RM (1997) Power tariff policy for groundwater regulation: efficiency, equity, and sustainability. Artha Vijnana, 34(3):312-322
Saleth RM (1999) Irrigation privatization in India: options, framework, and experience. Economic and Political Weekly 34(26):A86-A92
Saleth RM, Dinar A (1999) Evaluating water institutions and water sector performance, World Bank Technical Paper N° 447. World Bank, Washington, DC
Saleth RM, Dinar A (2000) Institutional changes in global water sector: trends, patterns, and implications. Water Policy 2(3):175-199
Saleth RM, Dinar A (2003) Water institutions and sector performance: a quantitative analysis with cross-country data. International Water Management Institute/World Bank, Colombo/Washington DC
Saleth RM Thangaraj M (1993) Distribution pattern of lift irrigation in India: an analysis by hydro-geological regions. Economic and Political Weekly 28(39):A102-A110
Salman SMA (2003) Inter-state water disputes in India: an analysis of the settlement process. Water Policy 4(3):223-237
Sangal SP (1991) Pricing of irrigation water in India. Economic and Political Weekly, 26(46):2645-2651
Sengupta N (1993) User-friendly irrigation designs. Sage Publications, New Delhi
Shah T (1993) Groundwater market and irrigation development: political economy and practical policy. Oxford University Press, Bombay
Shankar K (1992) Dynamics of groundwater irrigation. Segment Books, New Delhi
Siddiqui IA (1992) History of water laws in India. In: Singh C (ed) Water Law in India. Indian Law Institute, New Delhi
Singh C (1991) Water rights and principles of water resources management. In: NM Tripathi Private Ltd, Bombay
Singh C (1992) Water rights in India. In: Singh (ed) Water Law in India. Indian Law Institute, New Delhi
Singh C (ed) (1992) Water law in India. Indian Law Institute, New Delhi
Vaidyanathan A (1985) Water control institutions and agriculture: a comparative perspective. Indian Economic Review 20(1): 25-83
Vani MS (1992) Role of panchayat institutions in irrigation management. Indian Law Institute, New Delhi
Williamson OE (1975) Markets and hierarchies, analysis and antitrust implications: a study in the economics of internal organization. Free Press, New York
World Bank (1998) India water resources management sector review: initiating and sustaining water sector reforms, report No 18356-IN. World Bank, Washington DC
World Bank (2000) World Development Report-2000/2001: attacking poverty. Washington DC/New York: World Bank/Oxford University Press

Uphill Flow of Reform in China's Irrigation Districts

James E. Nickum

4.1 Introduction

From Beijing to Washington DC, the tocsins of water crisis in China have sounded (e.g., Ma 1999 or Brown and Halweil 1998). Homer-Dixon (1999 p 100) and Economy (1997) have raised the more distant possibility that if China's leaders cannot solve their country's proliferating water problems, state capacity itself may be at risk. While these concerns are legitimate, I have been and remain skeptical of the likelihood of a general or permanent water crisis (Nickum 1998). Historically, economies and institutions find ways to address critical water problems more often than not. Nonetheless, it would seem that in China, effective institutional change in the irrigation sector, evidenced by greater cost recovery, more efficient application rates, and user-driven water delivery, is likely to be a slow, long-term process.

The groundwork for change has been laid at the policy level, but the track record of institutional reform in China's water sector, especially for its state-operated irrigation districts indicates the presence of a number of factors, both internal but more importantly in the dis-enabling environment, that complicate and impede institutional reform in practice. It is not entirely clear, however, that the current system is failing. It depends on the criteria applied. By some standards, it may be all right for reform to be like molasses, slow but sweet. These may not be the standards we wish to espouse, however.

4.2 Locating the Irrigation Districts in Contemporary China

My focus here is on irrigation districts, a particular subset of water management institutions where inertia, both internal and external, appears to be high and seemingly intractable. In the history of the state-operated irrigation districts over the past 50 years, we find, in accelerated time, a complex interplay between intrinsic and extrinsic factors leading to an apparent secular decline in delivery capacity.

Under the People's Republic of China, established in late 1949, the area of irrigation has expanded significantly, especially in the dry north (Nickum 1995).

Construction of projects has usually involved some combination of state funding and materials and local resources, notably labour (Nickum 1977, 1978). Depending on project size, the management of facilities once they are constructed has in principle also involved a combination of state and local organizations, the latter divided into "democratic" and "mass" components. "Democratic management" institutions, which often have formal authority, have been exemplified by periodic gatherings of representatives of stakeholders (political, professional, user), and appear to have been about as democratic as most corporate shareholder meetings in capitalist societies (i.e., not democratic at all). "Mass" (now often termed "farmer" or "user") organization begins at the township or village level, and is responsible for basic-level watering operations and maintenance of "*peitao*" works at and below the tertiary (lateral) level of a canal system (*zhi qu*) (Nickum 1982:3).

The term *peitao*, often translated as "ancillaries," or "conveyance system," is a commonly used but rarely defined term[1] that appears to refer to all those works lying in the mass sector outside the direct administration of the state but within the command area. The interface between state-managed and local or user-managed irrigation is a problematic terrain in most of the world, especially where initiative to construct projects comes from the state.[2] China is no exception.

Peitao facilities and measures include not only the conveyance works from the lateral to the field, but on-field land formation (e.g., levelling and means of irrigation, such as flood, furrow, or pipe) and drainage networks where needed (Ministry of Water Resources 1994). Under appropriate conditions, such as areas with high water tables, it may also include the conjunctive use of wells and canals.

Direct state management is quite limited or altogether absent, not only in *peitao* areas but for most of China's irrigation, which is provided by facilities such as pumping stations and tubewells, with small coverage per project and relatively uncomplicated management structures (Table 4.1). This does not mean that these smaller projects are less of a problem when it comes to sustainability, however, due both to the public good (no excludability) nature of their underground sources and their dependence on electromechanical equipment. Ironically, although it is watered more efficiently than land under gravity systems, tubewell-watered land is far more likely to go out of irrigation. In particular, the nearly 4 million diesel or electrically powered deep tubewells that govern 15 million hectares concentrated

---

[1] For example, the massive and otherwise comprehensive three-volume *Water Conservancy Encyclopedia China* (1991) does not have an entry for *peitao*.

[2] "Project culture" (my term) can also lead to unintended adverse consequences with ecological as well as with human systems (Nickum 1999; Nickum and Greenstadt 1998). Recently, Chinese policy has called for a switch from "project water control" to "resource water control," reflecting recognition of this problem (Liu 2002; Wang Shucheng 2002).

in the north, especially in the provinces of Hebei and Henan, are prone to high rates of obsolescence and failure, due in part to widespread overdraft of aquifers.

Table 4.1. Effectively irrigated area (EIA) and irrigation districts, 2000

Province	Total [ha.]	ID [ha]	Large ID [ha]	TW [ha]
Beijing M	322,000	207,000	139,000	125,000
Tianjin M	353,000	77,000	0	112,000
Hebei	4,435,000	1,236,000	684,000	3,589,000
Shanxi	1,253,000	700,000	257,000	560,000
IMAR AR	2,372,000	1,264,000	952,000	1,155,000
Liaoning	1,441,000	521,000	291,000	630,000
Jilin	1,315,000	274,000	20,000	570,000
Heilongjiang	2,032,000	438,000	27,000	1,018,000
Shanghai M	286,000	0	0	0
Jiangsu	3,901,000	1,251,000	549,000	95,000
Zhejiang	1,403,000	699,000	315,000	25,000
Anhui	3,197,000	1,441,000	946,000	282,000
Fujian	940,000	298,000	29,000	18,000
Jiangxi	1,903,000	521,000	67,000	10,000
Shandong	4,825,000	2,210,000	1,051,000	2,347,000
Henan	4,725,000	1,222,000	687,000	3,296,000
Hubei	2,363,000	1,941,000	928,000	38,000
Hunan	2,667,000	1,065,000	324,000	0
Guangdong	1,981,000	749,000	157,000	21,000
Guangxi AR	1,511,000	654,000	93,000	4,000
Hainan	240,000	154,000	74,000	1,000
Chongqing M	625,000	94,000	0	0
Sichuan	2,469,000	1,241,000	715,000	40,000
Guizhou	653,000	97,000	0	0
Yunnan	1,403,000	540,000	179,000	10,000
Xizang AR	161,000	48,000	0	10,000
Shaanxi	1,311,000	876,000	452,000	450,000
Gansu	1,137,000	848,000	339,000	152,000
Qinghai	249,000	129,000	39,000	4,000
Ningxia AR	400,000	353,000	315,000	21,000
Xinjiang AR	3,128,000	3,344,000	1,694,000	544,000
TOTAL	55,013,000	24,492,000	11,323,000	15,127,000

Province includes 3 large municipalities [*M*] and 5 "autonomous regions" [*AR*] *EIA* effectively irrigated area *ID* irrigation district *TW* tubewell *IMAR* Inner Mongolian Autonomous Region

Source: Yearbook of China Water Resources 2001 (2001):416-417, 424

The institutionally more complex and extensive state irrigation districts rely predominantly on gravity delivery from impoundments such as reservoirs or diversions from rivers. Irrigation districts have been the primary focus of attention of an extensive literature published by central government organs such as the

Ministry of Water Resources. An example of this literature is a history of irrigation management (Ministry of Water Resources 1999) that is the basis for the next section. Because of this higher level of state involvement as well as the complexity of irrigation districts, information on their management is more readily available to the outside observer than for the smaller projects.

The history of management of irrigation districts indicates the importance both of the larger constitutional framework and of intrinsic institutional problems as factors leading to a diminishing of capacity over time. These problems are perhaps particularly visible in the larger irrigation districts, those defined as having an effectively irrigated area (EIA)[3] in excess of 20,000 ha.

While irrigation districts cover less than half the EIA in China, they are nonetheless quite significant. Actually, as table 4.1 shows, they are the most widespread single type of irrigation management form, covering 45 percent of the country's total EIA of 55 million hectares. The 242 large irrigation districts[4], averaging 44,694 ha each, govern nearly half of this area (11.3 million hectares). They are commodity grain and cotton bases for the state, generating about one-fifth of the country's grain (Yao 2002). The 5,441 small and medium-sized irrigation districts cover a slightly larger 13.2 million hectares (average 2426 ha).

4.3 Recent History and the Disenabling Environment

After its establishment, the People's Republic of China adopted a modified Stalinist economic model. This involved the nationalization of major industries and collectivization of agriculture; replacing the market mechanism with production-oriented planning; and extracting economic surplus from the rural economy by means of controlling the terms of trade ("scissors") between the agricultural and industrial sectors. Specifically, the scissors involved forced state procurements of grain for urban populations at artificially low prices, while the planned economy focused on increasing grain output through increasing inputs such as labour and water. This system has a built-in bias towards construction over operational efficiency, discourages the use of economic charges for natural resources such as water, inhibits technological change, relies on top-down decision-making operating through geographically-based "self-reliant" administrative units, and

---

[3] "Effectively irrigated area," the most commonly used measure of irrigation in China, refers to level land with water sources and complete facilities to provide adequate water "under normal conditions" (Nickum 1995).

[4] The number of irrigation districts changes regularly but not dramatically. The Minister of Water Resources provided a figure of 247 a few months after the source used here (Wang Shucheng 2003).

tends to have serious coordination problems at the interface of state and collective sectors (in China, usually the township level).

To put it mildly, efficiency is not a major characteristic of the Stalinist approach. Rigidity[5] is, and was one of the major reasons for its nearly worldwide collapse in the 1980s.

The period from about 1956 to 1960 witnessed great mass mobilizations of rural labour for the construction of water projects, linked closely to the reorganization of the rural areas into collectives of ever-increasing size, culminating in the monstrous but short-lived people's communes of the Great Leap Forward of 1958-1960. A significant proportion of China's surface storage and diversion capacity was created in this period, often to dubious standards of quality.[6] Over 900 of China's medium and large reservoirs, nearly one-third of the total, were built during the two years of the Great Leap Forward alone (Yearbook of China Water Resources 2001:478).

Almost from the beginning, concerns were expressed about poor management, especially at the field level, evidenced, for example, in low water use efficiencies, a lack of "planned use," inadequate *peitao* works to provide for effective and efficient application of water at the farm and field level, weak linkages between agricultural and irrigation measures, inadequately trained specialized personnel, and problems in water fee collection. In the early 1960s, focus shifted away from the construction of large-scale surface gravity systems to their "consolidation and readjustment". In the most extreme case, major diversions from the Yellow River into the flat plains of Hebei, Henan and Shandong were suspended in 1962 because of these weaknesses and consequent widespread rapid secondary salinization of the soil. In 1964 the Ministry of Water Resources announced that over 40 percent of the 19 million ha designed command area of large and medium irrigation districts had no *peitao* and were not operating properly. The total irrigated area continued to increase, but largely due to the spread of smaller pumping systems, at first on the surface in southern areas (Nickum 2003). At the same time, many smaller projects had "no one to manage them."

Beginning in late 1966, the tentative irrigation management structure that had been built up in the brief aftermath of the Great Leap Forward was ripped down

---

5 A rigidity, or delayed response to change factors, that I have elsewhere termed institutional "hysteresis" (Nickum 1994a).

6 China had almost no major dams in 1950. By 2000, the country had 420 large reservoirs, 2,700 medium ones, and 82,000 small ones. Size category is based on design storage capacity, not height of dam (large: over 100,000,000 cum; medium: between 10,000,000 and 100,000,000 cum; small: between 100,000 and 10,000,000 cum).

by the Cultural Revolution, which if anything exhibited a greater antipathy towards professionalism and economic measures than had the Great Leap. The winter mass mobilizations of rural labour for "farmland capital construction" were revved up again, involving as many as 100 million farmers (Nickum 1977, 1978). Many professional managers from the central government to the county level were "sent down" to basic production units, usually in the countryside. Those who remained on duty often did well to open and close the gates to release the water, and were naturally reluctant to impose or enforce a watering schedule in the midst of politicised chaos. Ironically, total irrigated area jumped by half between 1965 and 1975, due primarily to the spread of powered tubewells on the previously unwatered portions of the North China plain (Nickum 2003).

The water sector was one of the first to seek to restore order. A national "experience-exchange" conference convened in mid-1971 criticized the tendencies to "emphasize construction over management, emphasize core works over *peitao*, and to emphasize [the building of] projects over results." It also noted an unhealthy propensity towards creating single-use (usually irrigation) projects and towards neglecting soil conservation around facilities, as well as a tendency to rely on state subsidies even for basic operations and maintenance ("simple reproduction").

The 1970s are now considered to be an extension of the Cultural Revolution, characterized by political battles that only resulted in a clear victory for the pro-market reform forces at the Third Plenum of the Eleventh Communist Party in the waning days of 1979. At the time, however, there was already a great deal of concern with strengthening the management side of irrigation, through measures such as improved professional staffing and training, creating good linkages between professional and mass organs, and promoting financial independence through the collection of water fees, preferably (but rarely) on a volumetric basis.[7]

Once the Cultural Revolution was officially over, it was immediately made the scapegoat for various kinds of enduring management failures, including the lack of *peitao* construction, continued incompleteness in administrative structures, and weak technical capacity. Financial self-sufficiency and the importance of focusing on management over construction were emphasized, perhaps in tune with a rapid decline in the level of state support for water project construction. As we shall see, these maladies have persisted over time in a variety of policy environments.

At the same time, the situation became more complicated at the mass level in the early 1980s as the commune system was broken up and land came to be farmed on a household basis under a contract system. The subsequent semi-privatization and, even more, the chaos of system transition, led in many

---

[7] I provided a review of available literature on model irrigation districts published in the early 1970s in Nickum (1982), elaborating on many of these themes.

cases to the filling in of canals, looting of public equipment and even construction materials such as the bricks in power houses[8], and often a return to a traditional conflict-ridden method of each household being responsible for watering its own fields.

In 1985, in keeping with similar moves at the time to reform state owned enterprises in industry, the Ministry of Water Resources issued a comprehensive set of 12 "opinions on reform" that were aimed at turning the irrigation districts into viable economic "enterprises" with independent accounting, and responsible for their own profits and losses (Chen 1987). Emphasis was placed on the need to collect water fees based on the cost of delivering the water, with the specific standards to be determined by the provincial pricing bureaus, not by the individual irrigation districts. Later that year, the State Council issued a document setting out common principles for assessing, collecting and managing water fees. Water fee collection began to increase after that in many irrigation districts, but rarely by enough to cover costs and with considerable leakage to more profitable non-water uses (Turner and Nickum 1995). I will come back to this issue of water fees and revenue collection.

Other areas addressed in the Ministry's 1985 "Opinions" included a call to diversify sources of revenue beyond water fees; to sharpen incentives in the state (professional) sector by moving from guaranteed employment to a pay-for-service contract system, while providing more training and consideration of living conditions to staff; to turn over management from the state to the "democratic" stakeholders' associations; and to set up and strengthen local "mass" management groups. These initiatives, especially the latter two, strongly influenced the agenda for subsequent attempts at institutional reform, lasting up to the present.

It quickly became clear that any reforms would have to address a number of accumulated problems that were becoming increasingly salient, especially in North China. These included a drop in the available water source, due to both climatic change and increases in demand, and a decline in the capacity of projects, especially those built in the 1950s and 1960s, due to aging and prolonged neglect, itself a reflection of inadequate revenue generation.

A Ministry of Water Resources survey at the beginning of the 1990s found that 40 percent of the structures in the canal systems of 195 large irrigation districts surveyed were functioning poorly due to aging (including, presumably, from deferred maintenance). It was estimated that it would require 18 billion yuan at the current prices to renovate all the surveyed irrigation districts. By comparison, the national capital construction investment budget for water conservancy in 1995 was 14 billion yuan (8.8 billion yuan for new construction, 2.8 billion yuan for expan-

[8] Corroborated by my own field observations in Yucheng County, northwest Shandong, in 1984.

sion, and 2.5 billion yuan for replacements) (China Statistical Yearbook 1996:147). Some funding was subsequently provided, and financing methods became increasingly sophisticated, but the total available appears to have remained well below the estimated amount needed for the projects to recover from the depredations of aging, much less to extend their reach into improved on-farm *peitao*.

The 1990s and early 2000s have seen further efforts at reform, some of them quite significant, especially in areas such as water pricing, establishment of legislation and water rights systems to meet the demands and possibilities of the market system, and institutional reform. One of the most recent of these efforts is State Council Circular No. 45 (2002) of 17 September 2002, on rural water works reform.[9] At the same time, stresses on the water resource system have intensified, especially in the trail of rapid economic growth and the concomitant urbanization (for Beijing, see Nickum 1994b, forthcoming) and opening up of trade and investment to the outside world.

Most of the institutional problems identified in the 1980s or before persist. Understanding of their causes has improved, but easy solutions appear to remain beyond reach. Aging continues to be a major concern, and field level *peitao* facilities cover less than 20 percent of the irrigated area. Control structures are absent and canals and drains intermingle. If facilities were turned over in their entirely to water users for management, they would have little to work with (Feng 2003). Cost recovery rates remain insufficient nearly everywhere.

A recent delineation of the problems currently facing irrigation districts in the Yellow River basin includes: (1) a continuous decline in the amount of water available for use; (2) serious aging of 40 to 60 percent of structures and large losses in the canal system due to leakage; (3) poor condition of field delivery systems and methods of irrigating, leading to excessive levels of water application; and (4) institutional and policy flaws (China Irrigation and Drainage Development Center 2002). Specifically, regarding the fourth point:

> The management mechanisms of irrigation districts are backward, the agricultural water price system is irrational, and the investment mechanisms for project construction and maintenance and repair are far from perfect. In the large irrigation districts, there is no positive mechanism linking construction, management, operations, maintenance and develop-

[9] The main contents and measures of this institutional reform include: (1) clearly defining rights and responsibilities of managing agencies; (2) classifying water management agencies by the degree to which their charges are deemed for public goods or commercial uses of water; (3) promoting organizational restructuring; (4) separating management from maintenance; (5) improving the water charge system; (6) establishing strict financial controls; (7) providing for downsized staff; (8) providing preferential tax treatment for transferred staff members; (9) improving management of new water projects; (10) reforming the management of small rural water projects; and (11) strengthening environmental protection and safety management.

ment; the division of responsibility among public good and enterprise objectives are not clear; and project construction programmes, management operations programmes, and water price mechanisms are not synchronized, making it impossible to attain all the objectives of supplying water, saving water and controlling pollution. These irrigation districts cannot meet the demands of the socialist market economy.

4.4 Reform Initiatives

As indicated above, the reform agenda has become quite large (e.g., Wang Shucheng 2003). I will focus here on two recent initiatives: institutional reforms promoted by the World Bank, notably the self-financed (or managed) irrigation and drainage districts (SIDDs); and attempts to reform and rationalize the pricing system. Both initiatives illustrate the complex and entangled interplay between the disenabling environment and internal inertia.

4.4.1 SIDDs

China has increasingly turned to international financial organizations for assistance in the development of new water projects and in the reform of existing ones. These organizations brought with them experiences and discourse on promoting participatory irrigation management (PIM) or even direct turnover that, at least rhetorically, are not incompatible with China's long-stated principles of democratic and mass management. The problem, as in many other countries, appears to be in the execution.

Two projects in particular are cited in the Chinese literature (e.g., Ministry of Water Resources Rural Water Control Department 1999; and China Irrigation District Association 2001:12-13) as seminal in the introduction of international concepts of PIM into irrigation management. I had the honour of being the institutional economist on the prefeasibility teams of both of these projects, but hasten to add that I was only one of many voices in the final proposals. One of these was a technical assistance project funded by the Asian Development Bank in 1988-1989 on reform of irrigation management and cost recovery, involving a detailed survey of six different types of irrigation districts. This project resulted in the formulation of some proposals for incorporating PIM concepts into China.

The other, and far more significant one, was the World Bank's Yangtze River Water Resources Project, launched in 1995 to provide support to water projects in Hubei and Hunan provinces. With encouragement from the Bank's task manager, Richard Reidinger, the Chinese introduced the concept of "self-financed irrigation and drainage districts" (SIDDs) and made it an integral component of this and many subsequent World Bank funded water projects.

The basic idea of the SIDD is to strengthen water management at both the user and the main system levels, and to provide an effective interface between them by establishing water users' associations (WUAs) as legal persons at the "retail" level that purchase their water "wholesale" on a volumetric basis from economically independent water supply corporations (WSCs), chartered as non-profit organizations (NPOs) under the Company Law. The WSCs were formerly under state management, usually as water resource "stations".

The SIDD combines experiences of the two provinces involved in the project. Hubei had been a lead province in creating experimental water user groups in the 1980s, and hosted a large International Conference on Irrigation Management Transfer in September 1994 (Johnson et al. 1995). These groups had strong farmer support and had yielded good results, but even they had some weaknesses from a PIM standpoint. The most important weakness was that their leaders were appointed by the government, not elected by the members, who saw the groups as appendages of the local Water Resource Bureaus (WRBs). Neither leaders nor members felt accountable to each other. Reflecting their rather formalistic nature, the groups were constituted and met once a year. Because of the top-down nature of their authority structure, the interface of the groups with the state sector did not enable them to address problems of main system management.

In Hunan, a water supply corporation was established at Tieshan to sell water provided by the reservoir funded by the project to the growing city of Yueyang as well as to WRBs, who presumably would pass the water down to the farmers via the administrative structure. The corporation did not seek to operate the irrigation system as a unified hydraulic unit independent of the administrative structure, nor to sell water directly to farmer WUAs – these concepts came with the SIDDs and the Bank project that supported them.

In an SIDD, WUAs buy water from the WSCs on a volumetric basis, measured at the headgate, which usually corresponds to the lateral. Sales agreements specify the rights and responsibilities of both parties. By the end of the year 2002, about 770 WUAs and 41 WSCs had been set up on a pilot basis in 13 provinces (including large municipalities), supported under World Bank projects (Reidinger 2001, 2002, personal communication 2003). By February 2003, there were 41 full SIDDS, including both WSCs or comparable water service organizations (WSOs), under World Bank projects (R Reidinger, interview 2003).

Some 2,500 WUAs have been established outside of World Bank projects, indicating spreading acceptance of this part of the SIDD model. WUAs follow hydrological boundaries, whereas previous mass organization below the lateral tended to be based on township and village administrative lines. Deliveries to the WUA are charged volumetrically, encouraging more frugal use in the field than is the case of conventional area-based charges. Volumetric water charges have the additional benefit of providing an incentive for the WSCs to be more efficient, since each cubic meter of water wasted is revenue lost. This kind of high-powered incentive is virtually impossible to stimulate in a government bureau.

On the other side, meter-based charges protect the user from being charged for water that is not delivered in appropriate quantities or at the right time. Area-based charges insulate the supplier from accountability for proper delivery.

The WUAs have considerable autonomy from the government. World Bank projects stipulate that this includes the right over the use of obligatory levies on rural labour (*yiwu laodong*) of up to 20 days unpaid work per year (Reidinger 2001). Water is delivered to the field by the association, mitigating the inevitable clashes, sometimes physical, between farmers when they are individually responsible for watering their fields. This reduction in conflict apparently results in a great improvement in the condition of vulnerable populations such as the handicapped and female heads of households, who are unable to fight for their water.

WUAs are widely accepted, except by some township governments, for whom the self-financing provisions constitute a loss of divertible revenue. Interestingly, quite a few township leaders are quite strongly supportive of the WUAs, because they recognize their larger value to the farmers and the township (R Reidinger, interview and personal communication 2003). Nonetheless, there is widespread concern that the WUAs are often only formalistically autonomous from the government and Communist Party hierarchies,[10] and that their independence will vanish once World Bank funding ends. The limited extension of the SIDD model outside of World Bank projects does give cause for concern in this regard.

Of perhaps even greater concern are the WSCs, which have been much harder to set up and, once established, tend to operate in a hostile institutional environment. Ministry of Finance regulations exempt them from most corporate taxes, since they serve agriculture, but local tax authorities nonetheless attempt to collect the exempted revenues. Where the local government supports the WSCs, this problem does not arise, and the WSCs produce good results efficiently. Hunan Province is actually using the WSCs as a vehicle to implement government policy reforms aimed at separating policy from operational responsibilities in government in Yueyang. There policy is to be handled by a new Water Affairs Bureau while full operational responsibility is devolved to the WSC. If others follow suit, WSCs could very well become a key mechanism for reforming the water sector (Reidinger, personal communication).

[10] Liu Bin (2002) notes a tendency for village governments to convert WUAs into an additional nameplate on the same "horse and rider," depriving them of autonomy. A similar situation is found with non-governmental organizations (NGOs). Most NGOs in China are "GONGOs" (government organized non-governmental organizations), located in government offices, with governmental officials serving as their officials.

Another problem facing WSCs serving agriculture is that they are not economically viable on their own under current rules. According to current pricing regulations, water to agriculture has public good characteristics and is not to be charged at full cost. The possibility of cutting costs to meet revenues better meets with little enthusiasm, as it implies reductions in the number of staff (Reidinger 2001). Concerns about financial viability, especially over possible negative impacts on the willingness of local governments to continue to provide their already limited subsidies for irrigation operations and maintenance, led to a change in the term for SIDD from "self-financed irrigation and drainage districts" to "self-managed irrigation and drainage districts" (Liu 2002; Reidinger, personal communication 2003).

4.4.2 Funding

One of the principal drivers of reform efforts such as the SIDD is the recurrent and possibly intrinsic problem of inadequate cost recovery and related phenomena of wasteful applications of water. The solution to these problems does not lie in change at the policy level, which has become increasingly sophisticated in its espousal of market-like price formation mechanisms that adjust to changes in demand and supply conditions. The primary difficulties would appear to lie in the complex disenabling political and institutional framework in which the irrigation districts are embedded, and with the economic realities underlying irrigated agriculture in an increasingly open economy.

Water fees

The State Council issued regulations authorizing the collection of water fees as early as 1965, but these were largely ignored in the closely ensuing Cultural Revolution, when water charges were denounced as the dark doings of "capitalist roaders". Water charges did not make it back to the official agenda until after the economic reform era dawned at the end of 1979. In July 1985, the State Council issued a major document on the assessment and reform of water fees that was aimed at the calculation of the costs of water supply (including operation and maintenance, major repairs and depreciation) and the assessment of water fees accordingly. Further regulations promoting the assessment and collection of water fees were issued by the central government in 1994, 1997 and 2001 (Jiang 2003). Yet both assessment and collection rates remain low, at least from the perspective of the entities providing water. The situation is often more complex from the viewpoint of the putative beneficiaries of water delivery:

> The system for managing water supply to agriculture usually entails the state-owned water management agencies managing the canals at the tributary level and above, while management below that is done by the township and village [on administrative lines]. Corresponding to this, the government only sets water price standards for supply by the state-owned water management entity to the tributary outlet; after that no one manages

them. As a result, the government, concerned about burdens on the farmer, [sets] the water supply price from state-owned projects very low and operates at a loss, while in actuality the water fees borne by the farmers are not reduced. The government's accommodating policy goals are not realized, while intermediaries who tack on surcharges find the taste of water sweet (No. 2 Team 2002).

In June and July 2002, the Pricing Bureau of the State Planning Commission and the Ministry of Water Resources jointly organized a national survey of water pricing issues in representative large and medium water management units, primarily irrigation districts. The reports of the five investigation teams, published later that year in *China Water Resources*, reveal a great deal about the status of pricing and the limitations that continue to impede adequate cost recovery through water pricing.

Although conditions vary widely from province to province and project to project, the price of water to agriculture rarely even approaches the cost. In southern China (Hubei, Hunan, Guangxi, Guangdong and Hainan), where the investigation data are relatively complete, water charges averaged roughly 35 percent of cost, and the collection rate of those charges from sites investigated ranged from 10.3 percent to 86 percent, the median being about half (Jiang 2003).

Why has poor cost recovery proven to be so intractable? Jiang (2003) attributes this implementation failure to fierce resistance on the part of water users, a reluctance on the part of the higher levels of government to impose burdens on the farmers, the weak economic condition of farmers who are operating at a loss if the value of their labour is factored in, and the poor condition of water supply facilities, especially for metering. He also notes the lack of an effective mechanism to compensate agricultural users for transfers of water to growing higher-value uses, even though these transfers require increased investments in water-saving facilities. Others point to continuing predatory behaviour by local governments, including the allocation of excessive staff, such as retired military personnel, to irrigation districts (No. 2 Team 2002).

In addition to the add-on charges attached by township and village governments, water fees are often seen as a source of general revenue by local governments as high as the county level. The levies exacted by these administrations can be quite significant; in one case cited in the survey, the counties took 55 percent of water fee revenue going to a reservoir irrigation district in Henan Province (No. 2 Team 2002).

Reasons for this squeeze are numerous, including: (1) a monopoly on political power by local government and Party officials that they are naturally reluctant to cede to entities such as SIDDs or autonomous enterprises; (2) a lack of adequate alternative sources of revenue to provide collective services such as health and education; (3) the importance of bureaucratic entrepreneurship in mobilizing public resources for use in profitable economic enterprises such as the "township and village enterprises" that have been one of the main engines of rural economic growth (Gore 1998); and (4) higher collection rates by local governmental agencies, in part due to their monopoly over power, in part because of their greater density of staff at and below the township level, where state irrigation districts have little presence on the ground.

Because of the last reason, until recently water fees were often collected in grain as part of the state procurement levy. Although the revenue from this collection was to be earmarked for water, in practice much was retained by the collecting agency. In 1998 the State Council forbade grain departments from adding surcharges for collecting water fees, so they ceased to provide this service (No. 1 Team 2002).

There are currently at least five different modes of collecting water fees: indirectly, via county financial administrations; and directly, by the water management unit, from township or village governments, from the farm households, or from farmer WUAs (No 1 Team 2002). Each mode has different advantages and disadvantages in terms of rates of collection, transaction costs, and handling fees, with the WUAs clearly attractive in most regards.

Multiple operations

In addition to water fees, state water suppliers have been encouraged to engage in other economic activities besides water supply ("multiple operations") to help make ends meet. The income from these activities has provided the bulk of gross receipts in recent years, but the share actually cross-subsidizing water projects is insignificant. For example, in 1997, multiple operations provided 66 percent of the 117 billion yuan revenue to the state water control system, while water supply provided 9.9 percent and power supply brought in 24 percent. Of the income from multiple operations, only 622 million yuan, less than 10 percent of post-tax profits, went to water projects (Xu 1999:369).

State water administrations have not been the only ones encouraged to diversify into often unrelated economic activities in order to cover budgetary gaps. Most notorious has been the People's Liberation Army and its 15,000 commercial businesses (Johnson 2000:160). Using unrelated profit centres to cross-subsidize unprofitable public enterprises has predictable consequences, including corruption from the misuse of public authority for economic purposes, and a shifting of the most capable management's attention away from the unprofitable public service areas of activity that are the agencies' putative reasons for existence.

Subsidies

The *Water Industry Policy* (*Shuili chanye zhengce*) issued by the State Council in October 1997 provides for state support for water projects that have a strong public good nature, such as flood control, watershed maintenance, water source protection and key works for agricultural irrigation and drainage (Ke and Gu. nd). So far, however, no good channel has been provided to supply adequate and secure subsidies. China's entry into the World Trade Organization (WTO) may make it difficult to provide direct subsidies to water or agriculture (Jiang 2003), although China's major trading partners appear quite capable of finding ways to subsidize farmers.

4.5 Interpretation and Meanings

What can we glean out of this brief review of China's irrigation districts? There are clearly some intrinsic difficulties with operating large surface systems over long periods of time that China shares with most other major irrigating countries (Coward 1980; Postel 1993; Johnson et al. 1995). For example, systems age quickly because of the lack of a reliable mechanism to generate resources to provide proper upkeep but also because of poor initial construction and natural factors such as sedimentation and termites. Impoundments of the heavily silt-laden waters of much of north China are not expected to last for more than a few decades under the best of circumstances. Degradations of the watersheds only accelerate the process. In this light, a dam may be seen as similar economically to a mine or oilfield, as a temporary boom but not for eternity.

State-operated portions of the system are not well integrated with the watercourse, where actual deliveries may be uncertain, unfair and certainly unmetered. State and collective ownership show their vulnerability particularly during periods of major institutional change, when valuable components such as copper wiring, pumps and even construction materials go missing.

It is quite possible that in many cases the costs of maintenance, including the transaction costs of collecting adequate water charges, are simply too great to allow sustainability. Advances in metering and technology and prosperity in general may lower the financial costs. Improved metering and direct fee collection may serve to reduce the perception gap between state managers and water users regarding matters such as the amount and timing of water delivered and the total amount of fees collected. At present, they appear to see two different realities. Nonetheless, significant institutional reforms such as the SIDD that raise the level of trust between State and users are probably necessary to make a significant dent in the transaction costs.

Despite these intrinsic problems, it would seem that it is the policy and institutional environment, despite its many revolutionary changes over the past 50 years, that has tended to have an overwhelmingly negative effect on the sustainability of irrigation districts in China. It would be surprising if it were otherwise. The primary objective of the Stalinist model was to extract grain from the countryside at below-market prices, relying on massive increments of inputs such as water, labour and fertilizer rather than economic incentives to generate output. More recently, grain prices have edged above world market levels and water charges have become more acceptable in some circles, but local governments, with an insatiable appetite for revenues, have made it very difficult to create an independent and effective water charge system. In this case, the institutional problem may not be a lack of funds for water but a surfeit of leeches.

Finally, it must be noted that in terms of system-wide objectives, we cannot say that the irrigation sector has failed or even that it is failing. Certainly, irrigation districts have not performed all that well by criteria such as watercourse maintenance, technical efficiency of water delivery, cost recovery, or effective involvement by the end user. Further efforts are no doubt necessary to improve performance in all these areas in order to meet the challenges of this century.

Nonetheless, China's agricultural sector has prospered despite these deficiencies. Recently rural economic growth has slowed down to about 4 percent per annum, quite respectable against nearly any measure except China's urban economic growth. The secular decline in the international price for grain, coupled with China's entry into the WTO, brings into question the economic desirability of extending irrigation, or even of maintaining existing irrigation when it presses against the economic and ecological margins. Employment arguments are shaky, given the growing diversification of China's economy and the significant proportion of China's farmers who continue to rely on rainfed agriculture. At bottom, it seems that the greatest arguments for institutional reform are more fundamental than economic: the development of an effective civil society and the protection of the vulnerable, including the women who are left to tend the farm while their husbands work in the cities, and the environment.

4.6 Acknowledgments

Since my involvement in the work described here spans a quarter of a century, a proper set of acknowledgments would exceed the allowed word count for the entire chapter. They would certainly include my old friends Liu Changming, Liu Chunzhen and Ren Hongzun, and the editors of this volume. Special gratitude in helping me prepare this specific manuscript goes to Richard Reidinger, for his efforts to keep me honest, up to date, and accurate on the development of the SIDDs and the latest policy developments.

4.7 References

Brown LR and Halweil B (1998) China's water shortages could shake world food security. WorldWatch July/August:10-18

Chen SS (1987) The management of irrigation districts as enterprises (in Chinese). China Water and Power Press, Beijing

China Irrigation and Drainage Development Center (2002) Strategic research on water-saving restructuring of large irrigation districts in the Huanghe [Yellow River] basin (in Chinese). Yellow River Water Resources Press, Zhengzhou

China Irrigation District Association (ed) (2001) Participatory irrigation management (in Chinese). China Water and Power Press, Beijing

China Statistical Yearbook (1996). China Statistical Publishing House, Beijing

Coward EW (ed) (1980) Irrigation and agricultural development in Asia. Cornell University Press, Ithaca NY

Economy E (1997) The case study of China: Reforms and resources: the implications for state capacity in the PRC. American Academy of Arts and Sciences, Cambridge

Feng BQ (2003) Engage in water user participation in irrigation management to promote a virtuous cycle in irrigation districts (in Chinese). China Water Resources A, March, pp 52, 66

Gore L (1998) Market Communism: The institutional foundation of China's post-Mao hypergrowth. Oxford University Press, Hong Kong

Homer-Dixon TF (1999) Environment, scarcity, and violence. Princeton University Press, Princeton

Jiang WL (2003) The capacity to bear water charges in agriculture (in Chinese). China Water Resources A, June, pp 41-43

Johnson C (2000) Blowback. Little, Brown and Company, London

Johnson SH, Vermillion DL and Sagardoy JA (eds) (1995) Irrigation management transfer. Food and Agriculture Organization of the United Nations, Rome

Ke LD and Gu H (eds) (nd) A guide to reading the water enterprise policy (in Chinese). Ministry of Water Resources and China Association of Water Law, Beijing

Liu B (2002) Reflections on extending SIDDs in the comprehensive agricultural development project (in Chinese). China Water Resources, May, pp 49-50

Ma Jun (1999) The crisis of China's water resources (in Chinese). China Environmental Science Press, Beijing

Ministry of Water Resources Rural Water Control Department (1999) A brief history of on-farm water control in New China (in Chinese). China Water and Power Press, Beijing

Ministry of Water Resources Rural Water Control Department (1994) The handbook of irrigation management (in Chinese). China Water and Power Press, Beijing

Nickum JE (forthcoming) Water and sustainability in Asian megalopolises: The case of Beijing. In: Sassen S, Marcotullio P (eds) Human resource system challenge VII: Human settlement development. Eolss, Oxford

Nickum JE (2003) Irrigated area figures as bureaucratic construction of knowledge: The case of China. International Journal of Water Resources Development 19: 249-262

Nickum JE (1999) After the dam age is done: Social capital and eco-partnerships in urban watersheds: with focus on the Lake Biwa-Yodo River Basin. In: Inoguchi T, Newman E and Paoletto G (eds) Cities and the environment: New approaches for eco-societies. UNU Press, Tokyo, pp 140-160

Nickum JE (1998) Is China living on the water margin? China Quarterly N° 156: 880-898

Nickum JE (1995) Dam lies and other statistics: taking the measure of irrigation in China, 1931-91. East-West Center, Honolulu

Nickum JE (1994a) Sustainable institutions for sustainable development. Regional Development Dialogue 15: 173-183

Nickum JE (1994b) Beijing's maturing socialist water economy. In: Nickum JE, Easter KW (eds) Metropolitan water use conflicts in Asia and the Pacific. Westview, Boulder, pp 37-60

Nickum JE (1982) Irrigation management in China: A review of the literature. World Bank Staff Working Papers Number 545. World Bank, Washington DC

Nickum JE (1978) Labour accumulation and its role since the cultural revolution. Cambridge Journal of Economics 2:273-286

Nickum JE (1977) Farmland capital construction: landscaping China. Contemporary China 1: 19-23

Nickum JE and Greenstadt D (1998) Transacting a commons: The Lake Biwa Comprehensive Development Plan. In: Donahue J and Johnston B (eds) Water, culture and power: Local struggles in a global context. Island Press, Washington DC, pp 141-161

No. 1 Team, Baijia Water Management Unit Investigation (2002) Report on investigation of water prices in four provinces and one autonomous region in southern China (in Chinese). China Water Resources, September, pp 21-24

No. 2 Team, Baijia Water Management Unit Investigation (2002) Report on investigation of water prices in the five northwestern provinces plus Henan, Shanxi and Inner Mongolia (in Chinese). China Water Resources, November, pp 43-45

Postel S (1993) Water and agriculture. In: Gleick PH, Water in crisis: A guide to the world's freshwater resources. Oxford University Press, New York, pp 56-66

Reidinger R (2002) Participatory irrigation management reform: Self-managing irrigation and drainage districts in China. Presented to the sixth international seminar on participatory irrigation management, Beijing, April 21-26

Reidinger R (2001) Water saving through participatory irrigation management reform: Self-financing irrigation and drainage districts in China. Paper presented to the International Seminar on Water Saving Agriculture, Beijing, in January, by JE Nickum on behalf of Dr Reidinger

Turner JL and Nickum JE (1995) Trickle down? Decentralization of water resource administration and financing in post-Mao China. In: Moench M (ed) Groundwater law: The growing debate. VIKSAT, Ahmedabad, pp 143-163

Wang SC (2003) Strive to promote sustainable use of water resources in order to contribute to the full-scale building of a prosperous society (in Chinese). China Water Resources A, January, pp 8-18

Wang SC (2002) The essential characteristics, theoretical basis and institutional safeguards for resource water control (in Chinese). China Water Resources, November, pp 6-8

Water Conservancy Encyclopedia China (1991) (in Chinese). Vol 3. China Water and Power Press, Beijing

Xu BP (ed) (1999) Twenty years of water business management (in Chinese). China Water and Power Press, Beijing

Yao G (2002). Reform promotes reformation, water-saving promotes efficiency (in Chinese). China Water Resources, September: 42-44

Yearbook of China Water Resources 2001 (2001) (in Chinese). China Water and Power Press, Beijing

Institutions for Water Management in Mexico

Cecilia Tortajada and Nancy Contreras-Moreno

5.1 Introduction

Water management practices in Mexico have changed in recent years mainly as a response to the varying economic, social and environmental conditions of the country. Generally, the management of water resources has traditionally given priority to infrastructural development, with limited considerations of economic instruments and social and environmental factors. The past few decades have witnessed challenges imposed by increasing demands from the different uses and users, as well as growing water quantity and quality concerns. As a result, decision-makers are being forced to consider not only technical solutions when dealing with water planning, management and development, but also have started to consider issues as appropriate institutional frameworks, implementation of demand management practices, protection of natural resources and the environment, participation from the affected sectors and stakeholders, formulation and implementation of legal frameworks, and improving management capacity.

Unlike in most other countries of the world, institutional arrangements for water resources management at the river basin level were first put in place in Mexico as early as the decade of the 1940s, with the primary objective to foster socio-economic development of the several regions based on water availability. These institutions, known as river basin commissions, operated for almost 40 years. When their importance vanished, new institutional arrangements, policies and instruments were developed according to the then priorities of the country. Later, during the 1990s, institutions were established again at the river basin level with the objective of improving water management practices. River basin councils were expected to work at the river basin level, and river basin commissions, committees, and groundwater technical committees at the sub-basin, micro-basin and aquifer levels, respectively.

This chapter analyses the challenges, achievements and constraints of the institution arrangements for water management in Mexico for the past six decades. It also analyses the reasons as to why, inspite of great efforts at the local levels, the achievements of the several institutions that have been established thus far have been limited in terms of rational and efficient management of water resources at the river basin level.

5.2 River Basin Commissions, 1946-1986

During the twentieth century, the economic and social goals at the national level predetermined what was expected to be the most appropriate policies and institutions "in order to cover the needs of the population and encourage national and local economic growth, taking into consideration the environmental capacities of the several regions of the country." (CNA 1995).

It was towards the end of the 1940s that Mexico embarked on large-scale water-based integrated development programmes[1] for the arid plains of the north, and the tropical areas in the east and southeast of the country. The then administration (1946-1952) focused on the use of water for the integrated development of the areas, with main emphasis in the tropical areas. There was a feeling that the tropics were backward areas where "everything had to be done", and there was tremendous optimism in the sense that development of these regions was a real possibility because of the availability of the untapped natural resources, mainly water (Estrada 2003). Water was recognised as an engine to trigger integrated development of the individual regions, through which the quality of life of the local people would be improved over both the short and the long terms, and the economic growth of the country would accelerate.

Specific river basins were selected where agricultural (including agro-industries at the commercial level), forestry and industrial development-related activities could be encouraged. A main objective was also to settle the different regions with population from the central part of the country where enough agricultural land was not available. People were encouraged to migrate to the tropics, and thus provide the labour that would be necessary for the development of the areas. The river basins selected had plentiful natural resources, but with population living in extreme poverty, with acute health problems and with inadequate social and support services in the sectors of health, education, communication or transportation.

In order to implement the water-based development plans, a strong institution responsible for water management at the central level was considered necessary. Hence, the then National Commission for Irrigation[2] was restructured to become

---

[1] Estrada (2003) has noted that objective of the integrated development programme for the river basins was to ensure that the different economic and social sectors would benefit from the natural resources available in the several basins.

[2] The National Irrigation Commission was created in 1926. Its first task was to construct infrastructure to irrigate the areas along the border with the United States to promote their economic and social development. At the national level, small irrigation projects were constructed for a more even geographical distribution of investments and accrued benefits. Irrigation districts and units were established throughout the country. By 1946, irrigation districts covered some 775,000 ha, and small irrigation projects accounted for more than 42,000 ha.

the Ministry of Water Resources. To coordinate the activities that would be carried out in the several basins, the river basin commissions were created as the first semi-autonomous, multi-purpose implementing agencies in the country. These commissions were responsible for the development and coordination of both plans and policies at the local levels to promote the growth of the most backward areas. The plans developed were based mainly on economic considerations, but were designed to take into account the social aspects of the several regions as well as the natural resources available (SRH 1952).

Executing and planning agencies were created at the river basin level. The executing agencies included the Papaloapan and the Tepalcatepec River Basin Commissions (1947), the Fuerte River Basin Commission (1951) and the Grijalva River Basin Commission (1952). In order to include larger areas within the integrated development programmes, the Tepalcatepec Commission was wound up in 1960 and the Balsas River Basin Commission, which covered a much larger area, was created the same year. Under the leadership of the Ministry of Water Resources, the following ministries participated in the development of the river basins: Ministry of the Interior, Ministry of Finance and Public Credit, Naval Ministry, Ministry of Economic Issues, Ministry of Agriculture and Livestock, Ministry of Communications and Public Works, Ministry of Health, and Ministry of National Properties and Administrative Control.

In general, the tasks of the executing commissions included, but were not limited to, planning, design, coordination and construction of irrigation projects, flood control programmes and hydropower generation. The commissions were also responsible for expenditures on urban and rural developments, health and communication services (including navigation, ports, roads, trains, telegraph, telephone services, etc.). Planning agencies included the Lerma-Chapala-Santiago and the Panuco Commissions (Figure 5.1).

When the river basin commissions were established, the prevailing policy emphasized the importance of the integrated social and economic development of the regions based on the natural resources available, water being the main resource. The commissions had full authority to plan and execute programmes for integrated development under their direct supervision. They were also responsible for coordinating the activities of the several ministries within the river basin, for which they had limited authority but on which they were able to make inputs. Because the river basins included more than one state, the power of the commissions was higher than that of the states and municipalities. Thus, even though the river basin commissions had the authority to plan and implement the different tasks in coordination with the specific ministries, the fact that the commissions were above the states created tensions among the institutions over the years.

When the commissions were first established, they had full support from the President, and hence they were practically autonomous with almost no budgetary limitations. However, this situation changed with time, since the following administrations had their own views as to what should be the role, if any, of the river basin commissions within the overall economic strategy of the country.

Fig. 5.1. River Basin Commissions

It is important to note that the performance of the different river basin commissions varied according to the socio-economic and political conditions of Mexico at specific times. These conditions depended, and still depend, on each six-year presidential mandates, and hence, on the priorities and budgetary allocations of each administration. For example, in 1947, the central government considered that the water resources in several geographical areas could act as "engines" for development of the poorest regions of the country. Accordingly, it provided very significant financial support to the Ministry of Water Resources for the implementation of the projects. However, the subsequent administrations had different approaches, which were reflected in the activities authorized and the budgets allocated. Hence, the achievements of the commissions were not always of their own making, since they depended mostly on the decisions at the federal level on which they had limited or no control.

In fact, the records show that the projects implemented by the commissions generally focused on the development of water projects, without much consideration of the long-term needs of the population, or the rational use of the all the natural resources available which could have contributed to the long-term economic development of the region (Barkin and King 1986).

Following is a brief analysis of the main activities and performance of the Papaloapan, Tepalcatepec/Balsas, Fuerte and Grijalva Commissions. For a review on the Lerma-Chapala-Santiago Study Commission, see Wester et al. (2001).

5.2.1 Papaloapan River Basin Commission

The area of the Papaloapan river basin is 47,000 km^2, with a run-off of 44,476 million m^3, representing nearly 12 percent of the national run-off. Fifty percent of this river basin is in the states of Oaxaca (164 municipalities), Veracruz (71 municipalities) and Puebla (29 municipalities).

The Papaloapan River Basin Commission was created on 26 February 1947 as "a technical and administrative body, with the objectives to construct all works for flood control, irrigation, power generation, communication (water transport, ports, roads, railways, telegraphs and telephones) and urbanization in the area, as well as to decide on industrial, agricultural and settlement-related issues within the integrated development of the region." (DOF 1947a). The creation of this Commission was partially influenced by the fact that institutions were being created in different parts of the world looking for the integrated management of river basins, especially the Tennessee Valley Authority (TVA). According to Poleman (1964) and Barkin and King (1986), the similarity between the Papaloapan Commission and the TVA was that both of them were conceived as developmental authorities to promote the integrated development of the river basins, with a main focus on flood control activities. The main difference, however, was that the TVA was an autonomous institution, whereas the Papaloapan Commission was under the Ministry of Water Resources.

The background for the creation of the Papaloapan Commission goes back to the frequent floods of the Papaloapan river. The records show that in 1921 and 1944, the flood damages were very high, both in terms of human lives and economic losses in parts of the states of Veracruz and Puebla. Hence, in order to develop appropriate flood control measures, the then President of Mexico commissioned a study with the main objective to propose specific alternatives for the overall development of the Papaloapan river basin. The study explored the possibilities for constructing hydropower projects to produce electricity, which in turn would industrialise the rural areas. The report was finalised in 1945. It concluded that it was necessary to create a technical commission for the study of the overall Papaloapan river basin (SRH 1972; SARH 1990) and for the implementation of the appropriate projects.

When the Papaloapan Commission was established, its first president was the Minister of Water Resources. The board of directors included representatives from the Ministries of Finance and Public Credit, and National Properties and Administrative Control. The funds for the projects and running expenses were provided by the central government through the Ministry of Finance and Public Credit. The first task of the Commission was to carry out several studies for which specific development projects could be formulated. As a result of these studies, programmes were established in the areas of sanitation, flood control, hydropower generation, communications, and agricultural, industrial and urban development (SARH 1990).

During the 1946-1952 period, some of the projects implemented by the Commission included the construction of the large multipurpose Miguel Alemán dam (flood control, irrigation, hydropower generation, and drinking water supply); the establishment of Alemán city (eventually for 150,000 people), construction of levees, training of the Papaloapan river, irrigation projects, water supply and sanitation works, roads, eradication campaigns for malaria and intestinal diseases, and construction of schools, markets and gardens (Poleman 1964).

However, in December 1952, a new federal administration came into power for another six-year period. While the overall objectives of the Papaloapan Commission remained the same, fewer projects were implemented. During this administration, the Miguel Alemán multi-purpose dam was completed, levees were constructed and new irrigation projects were initiated. Roads were constructed and improved, hydropower generating plants were installed, health-related projects were implemented and schools were constructed. The budget allocated was higher than the previous administration, mainly because of the infrastructure that had to be constructed (Orive 1970).

A main constraint to the success of the projects was the poor implementation of the economic policies, for example, agricultural credits and marketing, to promote the growth of more profitable crops. The problem was that an integrated plan which considered the overall management of the region was missing, the financial institutions did not provide the appropriate support and trained staff was not available. For example, the Commission reported that from 1954 it had to provide credits to the farmers from its own budget, as a result of which the credits available were very limited (SRH 1958).

The political priorities changed with time, and the Papaloapan Commission as an institution was not considered as important to achieve the economic policies of the country, compared to when it was established. Its budget was drastically reduced and it could not perform its functions as the institution responsible for integrated activities in the basin. It also suspended the construction of large projects or its participation on health-related activities, education, agricultural research, construction and maintenance of roads and schools, etc. During the 1958-1964 administration, there were virtually no new funds for investments and most of the limited ones that were available were allocated for operation and maintenance activities and for the construction of few small projects for drinking water supply and flood control. Consequently, the staff was reduced to about one-third of what it was in 1956. Additionally, in 1962, the then administration considered that the several river basin commissions should not be responsible for works that were not strictly water-related. Consequently, the Papaloapan Commission handed over the responsibility for road construction to the Ministry of Public Works (Orive 1970; SARH 1990).

The budget allocated to the Papaloapan Commission under various six-year presidential administrations, over the period 1947-1961, are shown in Table 5.1.

Table 5.1. Budgets allocated to the Papaloapan Commission, 1947-1950

Periods	Budget (million pesos in 1972 constant prices)
1947-1952	270
1952-1958	592
1958-1964	172
1964-1970	202
Total	1,236

Source: SRH 1972

Of the total 1,236 million pesos (at 1972 constant prices), 41 percent was allocated for water projects, 24 percent for communications, 11 percent for social activities, 10 percent for administration, 4 percent for agricultural development, and 5 percent for equipment (Table 5.2).

Table 5.2. Budget of the Papaloapan Commission for different activities, 1947-1972

Activities	Percentage of Investment funds
Water works	41
Communications	24
Social activities	11
Administration	10
Agricultural and livestock development	4
Equipment	5
Others	5

Source: SRH 1972

The studies that the Commission carried out on hydrology, climate, geology, agriculture and anthropology were very useful. However, irrespective of the overall investments that were made by the Papaloapan Commission, there was no real evaluation of its overall achievements and the lessons that could be learnt from such a development activity. Official reports focused mainly on the descriptions of the implemented projects (SARH 1990), but there was hardly any serious assessment of the experiences resulting from the several development plans, or even specific activities, carried out by the Commission. Some of the few broad impacts noted by the Papaloapan Commission (SARH 1958) were the following:

- Flood control, which during the years 1952, 1956 and 1957 would have resulted in economic losses of more than 800 million pesos, not including human lives;

- Agricultural production in the basin, estimated at 290 million pesos before 1947, increased at a 8.4 percent annual rate;
- Cultivated areas increased from 223,000 ha in 1947 to 394,000 in 1955;
- Hydropower was generated;
- Sugar production increased by 150 percent between 1947 and 1958;
- Approximately 171,000 ha were irrigated.
- More than 2,000 km of roads were constructed; and
- Revenue of the Papaloapan river basin in terms of taxes collected represented 56 percent of the overall public investment in the basin.

Globally, the Papaloapan Commission was one of the first programmes of integrated river basin management. It was also one of the very first development programmes which focused on the tropics. However, inspite of the initial interest to develop the social and environmental potentials of the area, the official concern was almost exclusively on flood-related problems. In fact, the main accomplishments of the Commission were in terms of flood control and communications works, mainly due to the completion of the Alemán dam in the first case, and road construction in the second. The benefits of the Miguel Alemán dam reached many users in the urban centres in terms of electricity, large-scale farmers (mainly sugar cane production), and ranchers (livestock development). In addition, large-scale farmers and ranchers were able to realize better prices for their products because of the road networks constructed. In terms of the expenditures for rural development, most of it was used to build Alemán city. Some parks, streets and recreational facilities were also constructed in other communities (Poleman 1964).

The achievements in terms of education, health and urban improvements were very important for the area. For example, malaria and yellow-fever were eradicated, and schools were constructed and improved, which benefited about 40,000 students. Even then, however, thousands of young people did not have access to education. In fact, the Papaloapan Commission noted (SARH 1990) that there was such an increasing demand for education that it was simply not possible for the Commission and the Ministry of Education to meet this demand. Even by 1987, when the Papaloapan Commission was dissolved, educational level of the area remained 10 percent below the national average.

Additional activities of the Commission included clearing unoccupied land to prepare for the settlements of the farmers to whom the land would be given for cultivation. It is important to remember that one of the objectives of the Papaloapan Commission was to settle "virgin land" with farmers, who were not from the region. The main problem with this policy was that there were already two million people living in the area, and the existence of virgin land was more of a myth than a reality (Díaz-Cisneros 1974).

Some major constraints were that the Alemán dam was designed for irrigation purposes before studies were conducted on the needs and desirability of supplementary irrigation in the area; Miguel Alemán city was laid out even though there were well-established towns near by; and a three-lane paved highway was constructed when there were not even unpaved roads. In addition, most of the settlement projects that were implemented ended in failure because of the lack of services available to the population (Poleman 1964).

While the policies of the 1946-1952 presidential administration focused on the construction of large projects, the 1952-1958 government chose instead more direct forms of assistance to promote development in the basin, such as secondary and feeder road systems and programmes for agricultural credit and settlement schemes. The general objectives of the new Administration were the intensive use of large land areas, both virgin and under use, and the improvements in educational and sanitary conditions. Contrary to what had been planned earlier, the budget for water projects, drainage and road construction was drastically reduced (Poleman 1964).

Overall, the agricultural development activities resulted in a few concrete achievements, partly due to the changing priorities and interests of the different administrations. For example, it was planned to construct a very costly gravity-flow canal from the Miguel Alemán dam to irrigate an area of 160,000 ha. However, this canal was too expensive and was thus never constructed. Two smaller irrigation projects were planned, Los Naranjos and Blanco river. Los Naranjos was supposed to irrigate a maximum of 4,000 ha and the Blanco river project, approximately 35,000 ha. In the case of Los Naranjos, the scheme proved to be uneconomical after two seasons of small-scale operation and was finally abandoned in 1957. The Blanco river scheme was never started because the local ranchers did not want to divide their land into smaller lots and refused to cooperate with the Commission. Orive (1970) has noted that under this project, only 1,800 ha were cultivated from the total that was planned.

In terms of the changes in the region due to the projects implemented by the Commission up to 1962, Poleman (1964, p 99) noted that "...the 16 years of the Papaloapan project have brought about important changes in the Papaloapan basin and its economy... If this suggests that the lower basin at least is now well on the way of sustained economic growth, it does not follow that the project has been completely successful. Nor does it follow that the Commission's activities have been based on the logical unfolding of a preconceived plan of development. Instead, much of the programme has evolved through trial and error, and on several occasions it has undergone major alterations in order to conform with changing national policies. The result has been a number of abandoned schemes, failures and expensive errors."

In general, settlement projects and credit promotion were not very effective. In many cases, they can be considered to be total failures (Wionczik 1982). Loans were available only to very few farmers, and the settlements were very small, with no more than 550 families. Consequently, in 1957, the funding for the Commission was drastically reduced (Poleman 1964). A very regrettable result was the displacement and improper resettlement of 22,000 mazatecos indigenous people due to the construction of the Alemán dam, where approximately 40 percent of the people spoke only their native languages and 56 percent could not afford any shoes. In this community, where the only economic activity was subsistence agriculture, about half of the families were resettled (McMahon 1989). According to Bartolomé and Barabas (1990), even by 1972, the indigenous people were still suffering from inappropriate housing, lack of water and electricity, and under the best conditions were connected through small roads which soon became unusable due to lack of maintenance. Even worse, most of the population did not receive title deeds to their lands, and thus could not obtain any credit. This indicates that the Papaloapan Commission, while entrusted to implement the objectives of the national policies which focused on agricultural and industrial development for which infrastructure had to be developed, totally failed to realize that more efforts were necessary to properly settle thousands of indigenous people who did not receive adequate compensations, and were mostly ignored by the development activities[3].

In relation to the conditions under which the Commission operated during the 1970-1982 period, Estrada (2003) noted that in 1970, with the 1970-1976 administration, regional development activities[4] were promoted as one of the main economic strategies to reduce the inequalities among the different regions of the country. The programmes implemented were for rural, agricultural and industrial development at the regional levels.

From 1972, the changes in the economic policies of the country, and hence water policies, resulted in radical modifications of the institutional frameworks of the country, including those related to water planning and management. The river basin commissions did not disappear, but were restructured to fit the new national development policies, mainly because these policies encouraged water planning at the regional level (SRH 1975). The programmes to develop the tropical areas were still considered important within the proposed development plans. Accordingly, the commissions found an appropriate niche for their work.

---

[3] More recently, Murillo (2003) carried out an overall assessment of the Papaloapan river basin, including the living conditions of the mazatecos indigenous people who were resettled. The mazatecos said that promises that were made to them when they were resettled have not yet been fulfilled. These included lack of paved roads and no drainage in several towns like Nuevo Ixcatlán, Nuevo Cantón and Las Pochotas.

[4] When the Papaloapan commission was first established, its objective was to achieve the integrated development of the basin. However, during the 1960s and 1970s, the official policies changed, and the focus changed to "regional development" with the objective to reduce inequality among the various regions. Integrated development was no longer considered to be important.

In 1972, a government commission was established to evaluate the economic development of the High Papaloapan river basin, including the demands from the population in terms of infrastructure. The recommendations stressed that the Papaloapan Commission should receive further financial support and become the agency responsible for regional planning as well as for irrigation and livestock-related activities, aquaculture, credit-related issues and construction of all necessary infrastructure (SARH 1988). The Commission became again the coordinating agency for the activities carried out by all the ministries within the basin.

With the beginning of the construction of the Cerro de Oro dam, the budget of the Commission increased by 117 percent. However, approximately 26,000 Chinantecos indigenous people were affected by this dam, and their resettlement process was not only inadequate but was also totally unsatisfactory. A main reason for this sad state of affairs was because the Papaloapan Commission did not pay enough attention to the demography and ethnic characteristics of the population: the emphasis was primarily on the number of persons who had the title deeds (Bartolomé and Barabas 1990)[5].

In 1974, the Ministry of Water Resources and the Papaloapan Commission prepared jointly a project, the objective of which was the integrated regional development for the overall river basins of the country. The Inter-American Development Bank provided a credit and the Papaloapan Commission, acting as an administrative and not as a coordinating body, engaged in rural development activities to increase agricultural production by organizing the farmers, and providing technical assistance and credit (SRH 1988). The Commission was responsible for planning and development. It also provided funds to other organizations to carry out their projects, which included irrigation works, drinking water supply and sanitation projects, electrification, and construction of roads, schools and clinics.

Figure 5.2 shows the annual budget allocations of the Papaloapan Commission from 1947-1985. It indicates extremely high variations in the budget from one year to another, which made any medium- to long-term planning difficult, if not impossible.

[5] In his assessment of the Papaloapan river basin, Murillo (2003) has noted that the chinantecos indigenous people are still lacking services which were offered to them such as paved roads, sidewalks and drainage. In addition, for example, in Las Pochotas, people complained that they were not given the 20 ha of land they were promised, but only 8.5 ha; and that in Los Naranjos, there are no irrigation activities or roads.

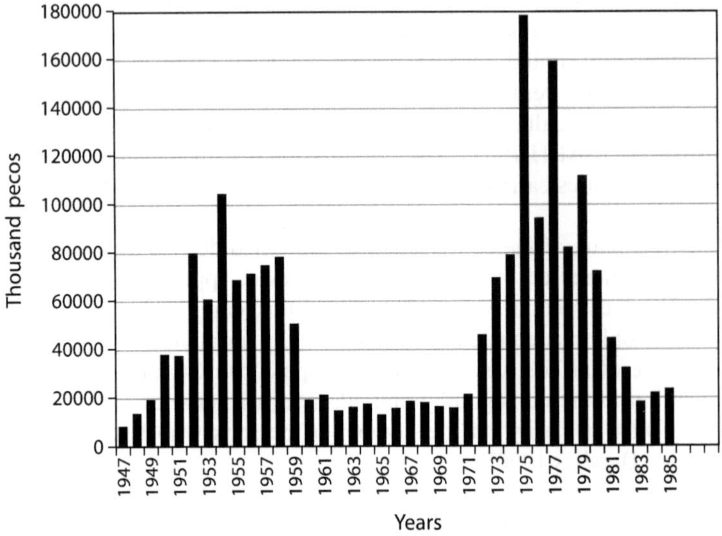

Source: SARH (1990) in Estrada (2003)

Fig. 5.2. Annual budgets of the Papaloapan River Basin Commission, 1947-1985

In 1982, the economic crisis affected negatively the whole country, including all national and regional level institutions. During this time of economic crisis and changing development policies priorities, the then administration decided to dissolve the river basin commissions. The Papaloapan Commission was officially dissolved on 24 December 1986 (DOF 1986), and projects and activities were handed over to the state offices of the SARH.

5.2.2 Tepalcatepec and Balsas River Basin Commissions

The second river basin commission that was established was the Tepalcatepec Commission, created by a presidential decree on 19 December 1947. It followed the same strategy as the Papaloapan Commission. Its main objectives were "to plan, design and construct the infrastructure necessary to develop the areas within the Tepalcatepec river basin" in an integrated manner. The Tepalcatepec river basin covers some 18,000 km^2 in the states of Jalisco and Michoacán. Even though this basin was not affected by severe floods, there were acute social, health and communication-related problems that the Commission was expected to address (DOF 1947b).

As in the case of the Papaloapan Commission, the president of the Tepalcatepec Commission was the Minister for Water Resources, and the board of directors included representatives from the Ministries of Finance and Public Credit, and National Properties and Administrative Control. The funds for its operation were channelled through the Ministry of Finance and Public Credit.

The main tasks of the Commission were the construction of irrigation networks within the Tepalcatepec river basin, rehabilitation of irrigation networks to improve their efficiencies, and the construction of the necessary road infrastructure to market the products that would be produced throughout the region (DOF 1947b). Extensive technical studies were carried out on issues such as population, education, health, sanitation, water resources, soils, planning of irrigation and hydropower projects, credits, urban and rural electrification, communication, forests, etc. (SRH 1961). The area originally selected to be under the Tepalcatepec Commission included three sub-regions of the state of Michoacán: cold jagged mountain chains, mild temperature areas, and coastal-tropical areas.

During the 1947-1952 period, some of the projects constructed under the Commission included the Valle de Juárez, San Juanico, Jicalán, Cóbano, Coróndiro, Piedras Blancas and Punta del Agua dams; 90 km of main canals, and 3.6 km of tunnels, in addition to numerous small-scale irrigation works. Other activities included water supply and drainage projects, as well as provision of electricity for urban and rural areas, construction of main and secondary roads, construction of schools and, most importantly, the eradication of malaria, gastro-intestinal diseases and tuberculosis in the region (Orive 1970).

The annual budgets allocated to the Tepalcatepec Commission during the period 1947-1958 are shown in Table 5.3. As indicated, the budget increased from four million pesos in 1947 to 41 million pesos by 1952. It is worth noting that during 1952-1958, the investment for irrigation activities decreased overall in the country for the first time in 27 years.

As it can be observed, the budget for irrigation programmes of the Tepalcatepec Commission was reduced from 7.5 percent in 1953 to 4.9 percent in 1958. Similar to the Papaloapan Commission, the projects implemented during this period were fewer than before. Projects for irrigation and power generation were constructed, which allowed rural areas to receive electricity. By 1958, the main impacts of this Commission were considered to be a significant increase in literacy (a large number of schools were constructed in remote areas through which more than half of the illiterate population learnt to read and to write), and construction of road networks which connected isolated communities to the rest of the country. The economic benefits were calculated to be around 400 million pesos, based mainly on the value of the agricultural production and the electricity generation (Orive 1970), in addition to the commercial and industrial activities that were promoted in the basin.

Table 5.3. Annual budgets of the Tepalcatepec Commission, 1974-1958

Year	Annual budget (Million pesos)	Percentage of total investment funds in irrigation programmes	Budget in 1996 pesos (Millions)
1947	4	1.8	11
1948	10	4.2	25
1949	13	5.2	30
1950	17	4.7	36
1951	21	4.9	35
1952	41	7.5	67
1953	43	7.5	72
1954	34	5.5	52
1955	32	5.0	44
1956	32	5.3	41
1957	32	5.0	39
1958	32	4.9	38
TOTAL	311	5.35	490

Source: Report of the Ministry of Water Resources (1958) in Orive (1970)

One of the major limitations of this project was that the Commission focused mainly on the development of the tropical areas without much concern for the mild- and cold-temperature zones, where most of the local farmers lived. In addition, the small farmers who lived in the tropical areas could not benefit from this development, except for irrigation in some cases. One of the reasons was that the farmers did not have access either to credit or to technical assistance and could not grow profitable crash crops like cotton, cantaloupe, watermelon or lemons. In other words, the development plan of the Tepalcatepec River Basin Commission failed in many cases to provide support to the farmers so that they could have access to credit and technical assistance. The credits available from the agricultural bank were insufficient, and were seldom provided on time. Under these circumstances, and as the years passed, the subsistence farmers did not have any other alternative but to rent their land to large-scale farmers who monopolised both the land and the credits available. In order to make a living, many of the farmers had to work as share-croppers for the big farmers who had rented their land (Díaz-Cisneros 1974).

Geographically, the Tepalcatepec river basin is within the Balsas river basin. Hence, in order to cover a larger geographical area, the Tepalcatepec Commission was dissolved, and the Balsas River Basin Commission was established on 11 November 1960. The action plan for the newly created Balsas Commission included the Federal District and the states of Michoacán, Guerrero, Jalisco, México, Tlaxcala, Puebla, Oaxaca and Morelos. The tasks of the Balsas Commission included the study, planning, design, construction and follow-up of all infrastructural development for flood control and irrigation, hydropower generation, and activities related to water supply, health, communication networks, trains, telegrams, telephones and ports. The Balsas Commission was also responsible for agricultural, industrial and livestock developments, as well as for human settlements (DOF, 1960). While the Tepalcatepec Commission was responsible for development projects in 17,915 km^2, the Balsas Commission was responsible for an area of 112,161 km^2. The investments funds allocated to the Tepalcatepec Commission increased from 21 million pesos in 1959 to 177 million pesos by 1964 (including 120 million of internal financing), after it became the Balsas Commission (Orive 1970).

The Executive Director of the Balsas Commission was a respected past-President, Lázaro Cárdenas, who managed to bring considerable prosperity to the basin. The activities of the Commission included mostly the planning of projects for hydropower generation, agricultural development, water supply, communication networks, etc. Irrigated areas were developed and roads and large dams were constructed. Water supply and drainage projects were implemented; improvement of urban areas included an average of 17 communities per year and the construction of 11 schools per year. The annual value of the agricultural production in the irrigation districts within the basin reached 735 million pesos in 1964 (Orive 1970).

As with the rest of the commissions, the Balsas Commission was dissolved after some 20 years of existence.

According to Barkin and King (1986), the main investment of the Tepalcatepec/Balsas Commission in the area was in terms of irrigation, drainage, roads and pilot-schemes for agricultural developments. Social projects represented only 10 percent of the total budget of the Commission until 1965. However, the Commission had very positive impacts on the people in the region. For example, the construction of schools was fundamental to promote education, even though this investment represented only 3 percent of the total expenditure of the Commission. Water supply, rural electrification and urban improvements also benefited the local populations, who contributed with labour so that as many towns as possible benefited with services at low cost. Access to electricity increased at twice the national rate between 1950 and 1960. In terms of health programmes, the Commission also contributed with limited funds to the eradication of malaria[6]. In addition, hospitals and clinics were constructed by the Commission for many communities. It should be noted that while the social projects were implemented for the entire basin, most of the economic investments were made mainly in the tropical areas. Overall, because the credit facilities were limited, only a few people could benefit from the prosperity brought about by the irrigation projects.

All the activities implemented by the different river basin commissions had beneficial impacts on the overall development of the country. The problem, however, was that the programmes and projects did not seem to have been successful in achieving cumulative effects at the regional levels. In the specific case of the

---

[6] This programme was promoted jointly by the governments of United States and Mexico.

Tepalcatepec Commission, perhaps the most important achievement for the development of the area was that it brought to the attention of the government the need for additional investment on a variety of development-related issues. Hence, while the spending on social issues by the commission may be considered to be limited compared to the needs of the region, the investments made by different institutions and communities were very important for the improvement of the quality of life of the population (Barkin and King 1986). This was an important contribution.

5.2.3 Fuerte River Basin Commission

The Fuerte River Basin Commission was created on 27 August 1951, as a "technical and administrative body under the Ministry of Agriculture with the objective of achieving the integrated development of the Fuerte river basin." As with the previous commissions, its objective was to foster economic development of the area where water could be the engine for development through the construction of infrastructure, agricultural and industrial developments, and hydropower generation. It was expected that agricultural development would result in extensive industrial development which, in turn, would result in the development of new urban areas, ports, airports, roads, trains and communication networks (DOF 1951).

In 1952, the Fuerte Commission took over the activities of the Fuerte irrigation district, which included planning and construction of projects for flood control, irrigation, hydropower, water supply, sanitation, and road networks and also improvement works. During its first year of operation, the budget of the Commission was four million pesos, which represented 0.7 percent of the total expenditure for irrigation at the national level. The budget increased later, especially during the construction of the Miguel Hidalgo y Costilla dam, when the figures represented 5.4 percent of the total investment in irrigation in the country in 1958 (Orive 1970). On 27 May 1958, it was agreed that the Fuerte River Basin Commission would also be responsible for the development works of the Sinaloa river basin, covering a total area of 35,000 km^2, with very rich land in the states of Sinaloa, Chihuahua, Sonora and Durango (SARH 1986).

During the 1958-1964 period, projects on irrigation, rural development, water supply and sanitation were constructed. Approximately three million pesos were invested per year in schools (55 percent), water supply (30 percent), electricity, roads and urbanization-related activities (15 percent). Overall, some of the projects started by this Commission included maintenance works for irrigation projects, technical assistance, water supply and some financial support for schools, electricity, roads and urbanization projects (Orive 1970).

In 1964, the objectives of the Commission were modified so that they would be compatible with the new regional water plans. The new priorities became planning, construction and operation of water projects, mainly to transfer water resources towards the northern areas where water for agriculture and livestock production had become important. Within the Water Plan for the Northwest of Mexico, the Josefa Ortiz de Domínguez dam was constructed during the 1964-1970 administration, and the Ing. Guillermo Blake dam during the 1970-1976 administration (SARH 1986).

The Fuerte River Commission was operational from 1951 to 1985. Following the changes in the national policies of the 1982-1984 period, it was decided that the Commission would become part of the government of the state of Sinaloa. The Commission was dissolved by a presidential decree on 8 May 1985 (DOF 1985).

5.2.4 Grijalva River Basin Commission

The Grijalva River Basin Commission was established in 1952, but it was not until 1959 that the president and its board of directors were appointed. The objective of this Commission was to promote the development of the southeastern regions of the country, including parts of the states of Tabasco and the Yucatán peninsula, incorporating approximately 50,000 km^2. The specific objectives included flood control, irrigation development, water supply, sanitation, and road construction (DOF 1951).

The first stage of the work of the Commission included a series of technical studies, extensive drainage works and construction of very essential roads. Flood control projects were constructed covering more than 260,000 ha. Since one of the main objectives of the Commission was to implement an extensive programme of settlement, with the support of irrigation and promotion of agricultural practices in tropical areas, the Chontalpa Plan was initiated during the 1952-1958 period (SRH no date). Additionally, the Malpaso dam was constructed between 1960 and 1964, and the Malpaso hydroelectric plant was built during 1964 to 1967.

The Chontalpa, in the coastal plain of Tabasco, was a mixture of forest (43 percent), pastures (24 percent), annual crops (22 percent) and perennial crops (11 percent). There were acute health problems because the only source of water available to the population was contaminated; most of the land was used for subsistence agriculture with very low returns and hardly any commercial value; and there were no roads for the distribution of agricultural or animal products. The Chontalpa Plan was meant to be a model for integrated development, where housing, education, health, irrigation, technical assistance, and communication were expected to be properly planned and implemented (Orive 1970). The Chontalpa Plan included the development of thousands of hectares of land in two phases. The first phase was to be implemented from 1965 to 1974, and included the development of 140,000 ha in the left bed of the Mezcalapa river. The second phase was to be implemented in the areas of Cuduacán, Jalpa and Nacajuca, but depended on the construction of infrastructure in the Samaria-Mecoacán areas (Contreras-Moreno 2002).

On 10 January 1972, the trusteeship for the Chontalpa Plan was established under the Ministry of Finance, and the Southeast Bank was to provide credits for agriculture and livestock development. There was an initial investment of 25 million pesos, out of the 200 million pesos that were to be invested during the first stage of the Chontalpa Plan. The participating institutions were the Ministry of Agriculture and Livestock, Ministry of Health, Administrative Committee for the Construction of Schools, Ministry of Education, Federal Power Commission, National Commission for Popular Subsistence, Southeast Bank for Agricultural and Livestock Development, National Bank for Agricultural Credit, and the government of the state of Tabasco.

In 1971, the Grijalva Commission was requested to plan and implement a second project for agricultural and livestock development in Tabasco, the Balancán-Tenosique Project. It was planned to be an integrated development plan under the same lines of the Chontalpa Plan, with an investment of 400 million pesos. The project covered some 115,000 ha of land, and was to be developed in two stages. The first stage included the development of the northern part of 50 000 ha (10 ejidos, or common lands, of 5,000 ha each). Some 20 parcels were established for agriculture, livestock and forestry development (Diaz-Cisneros 1974).

The Balancán-Tenosique Plan included the construction of main and secondary roads that would connect the communities. Pre-feasibility studies were carried out. It was found that 10 percent of the land was suitable for livestock production and the balance could be used for irrigated agriculture. Some 20 population centres were to be developed, with services such as water supply, sanitation, electricity, hospitals, primary schools, sports centres, etc. Some 250 families were expected to live in each centre. The land, each of 20 ha, would become cooperatives and land owners would work jointly to increase productivity, promote technical development and revenue generation, develop industry and generate employment. It was expected that the small-scale industries would be integrated, both vertically and horizontally, with the economic activities of the region, and that the added values would remain in the area and would benefit the farmers. During the planning stage of this plan, there were about 1,000 families from the state of Tabasco, Michoacán, Guanajuato and Chiapas in the region. About 4,000 families from Tabasco were expected to be settled in the area. The plan identified areas for the development of natural forests or plantations of valuable trees such as cedar and mahogany, as well as orchards of fruit trees. The economic activities would include agriculture, animal husbandry and bee-keeping. In addition, two areas were selected which would be under the control of the government for reforestation and research activities (Contreras-Moreno 2002).

In 1973, the Executive office of the Grijalva Commission assessed the performance of the Commission (Echeagaray 1973). According to this evaluation, up to 1973, the total investments of the Commission was approximately 2,360 million pesos: 118 million pesos during the 1952-1958 period; 1,026 million pesos during 1958-1964; 840 million pesos during 1964-1970; and 376 million pesos during

1972-1973. Out of the total amount, 980 million pesos had been invested in the Chontalpa Plan.

The assessment report concluded that the overall impacts of the Chontalpa Plan were very poor. The objective was to develop 140,000 ha during the first stage of the plan and 135,000 ha during the second stage, for a total of 275,000 ha. The first stage was also divided into two phases: a first phase of 80,320 ha, which was under development during the time of the assessment, and the second phase of 59,680 ha which was to be developed in the near future. In reality, out of the 80,320 ha that were being developed, works and services had been implemented to utilise only 26,700 ha for agriculture and livestock-related activities. The rest was "being wasted" because of lack of credit and absence of viable activities for the local population. The conclusion was that, 20 years after the plan was formulated and had received an investment of about 980 million pesos, it had not resulted in the expected outputs. The population did not benefit from the projects and services mainly because of lack of credits. Due to such poor results, it was recommended that highly profitable crops like banana, cacao and sugar cane would have to be introduced in the area, which might result in short-term profitable returns on the high investments that had already been made (Echeagaray 1973). According to Redcliff (1982) employment generated in the area was very limited and the peasants were ousted from their lands. The main constraint was that the control of the project was in the hands of financial institutions and the project was basically planned and implemented from Mexico City (Barkin and King 1986), with virtually very little local involvement.

The assessment by Echeagaray (1973) also pointed out that the health and education-related activities, which were the responsibility of the Grijalva Commission, were handed over to the Ministry of Health and the Ministry of Education in 1972. In addition, the activities related to trusteeships had also been handed over to the National Bank for Agricultural and Livestock Development. The assessment strongly recommended that the Commission should be restructured in order to become an implementing body for integrated development, since "at present, it has become basically a regional office for water resources for the state of Tabasco, with no decision-making power in the state of Chiapas, where the High Grijalva is located. It is necessary to develop an integrated plan for the development of the entire Grijalva basin" (Echeagaray 1973, p 5).

In his analysis of the Chontalpa Plan, Barkin (1978) noted that the plan was originally conceived as a pilot project for the management of the humid tropical areas of Mexico, which, if successful, could provide solutions to political and economic problems the country was facing. However, inadequately planned and implemented projects resulted in massive deforestation and ecological deterioration. The objectives of the plan could not be achieved because of lack of technical assistance to the farmers, inadequate availability of credits and unstable employment generation.

No records exist at present as to whether the Grijalva Commission was restructured, or if the Chontalpa Plan was modified to suit the interests of the region. The report of activities of the Grijalva Commission 1974-1975 (Grijalva Commission 1975) only notes some activities on drinking water and sanitation in urban and rural areas, irrigation, drainage and aquaculture, with investments of millions of pesos.

The Commission was dissolved in 1985, and its responsibilities were transferred to the state offices of the Ministry of Agriculture and Irrigation.

5.3 Factors which influenced the performance of the river basin commissions

In order to understand the water-based integrated and regional development programmes and institutional arrangements, it is necessary to analyse the forces that provided the context for these programmes and institutions. It is important to note that water and agricultural policies have been intertwined for decades as part of the economic development strategies of the country. Hence, any change in the water policies affected the agricultural policies, and vice versa.

In 1947, when the river basin commissions were established, the economic policies of the country focused primarily upon large-scale agricultural and industrial development projects. Numerous large projects were constructed for different purposes, the most important of which were for irrigation and hydropower generation. The investment in irrigated agriculture was extensive. From 1947 to 1967, irrigation was introduced in more than 1.2 million ha. The expectation was that expansion of irrigated agriculture would bring agro-industrial development, increase exports, and generate employment opportunities.

However, a major constraint for policy implementation for all the sectors, including those in water and agriculture, was that the programmes and projects at both national and local levels had to be planned and implemented only within the six-year presidential administrations. Historically, with the election of a new President, even though he was from the same political party as his predecessors, the new Administration invariably made radical changes in national priorities, policies and programmes. This has meant that the continuity of major development programmes and projects can seldom be assured beyond the six-year Presidential term.

Regarding the development of the rural sector, according to Schumacher (1981), de Janvry (1995), and Diaz-Cisneros (1974), interest and support in irrigated agriculture, and thus in water resources development, have varied dramatically within the different six-year presidential periods. For example, during the 1946-1952 presidential term, the national policies were to invest heavily in the different regions, more with economic than social objectives. The administration that followed during 1952-1958 faced high inflation rates and currency devaluations. Consequently, the total budgets of the different ministries, including the river basin commissions, were reduced in real terms. However, since there were large development projects under construction (mainly for irrigation and hydropower generation), the Ministry of Water Resources still received some priority. Inspite of this, its budget was reduced from 13.8 percent of the total national budget in 1953 to 7.8 percent of the total national budget in 1958. The support to the river basin commissions continued, but the projects implemented became fewer. More importantly, the power and authority of the commissions were steadily reduced. The political perception was that the commissions clashed with the other ministries and with the state governments. For example, the Tepalcatepec Commission found itself having no funds for the construction of new development projects, and the Fuerte Commission lost its initial mandate (Orive 1970). The 1952-1958 administration achieved higher agricultural outputs, but it was accompanied by high levels of rural under-employment.

At the beginning of the 1960s, national exports declined, investment plans for the public sector were changed, and the budget allocated to the agricultural sector was drastically reduced, impacting negatively on the performance of the river basin commissions. Less irrigated areas were developed but there was more financial support for the maintenance of the irrigation districts. From 1964, the water policies of the country emphasized not so much the construction of water projects, but the improvement and development of small irrigation projects primarily because the objective was to increase the social benefits. In addition, the concept of integrated development of basins was changed to integrated regional development, which included several basins. For water resources planning, several regions were established and water plans were formulated for the northwest, central and central-gulf regions. A National Plan for Drinking Water was prepared to ensure future availability of water supply for the Mexico City Metropolitan Area (SARH 1988).

In 1972, a new water law was enacted[7], and a commission was established to develop a national water plan, with national and regional objectives. In 1975, the National Water Plan was finalised, and in 1976, the National Water Planning Commission was established to implement it, and to update it on a regular basis (CNA 1995). The 1975 Water Master Plan stated that the management of water resources would be carried out at the regional level based on the hydrology of the country. The plan also proposed the establishment of water institutions at the regional level, which would include the major river basins. These institutions would have decision-making power and would have the authority to formulate the regional water development plans, prepare and implement actual projects, and decide on the fees and collect them for water users and effluent discharges. The central authority was to be responsible for the policy formulation at the national level, coordination among the different sectors, resolution of conflicts among the re-

[7] The previous water law was enacted in 1932.

gions, integration of regional plans into a national planning framework, implementation of large-scale and technically complex projects, and management of research and training programmes (SRH 1975).

According to the 1975 Water Master Plan, 13 hydrological regions were established, and regional programmes as well as water development programmes were developed for each one of the regions, with the objective of using more efficiently land and water resources available, and to reduce the prevailing inequalities in water availability to the people. The water development programmes included large-scale irrigation, flood control and drainage for agriculture; water supply for major urban and industrial areas, and multi-purpose hydropower projects (Herrera-Toledo 1997).

In 1976, due to the importance of irrigation at the national level, the Ministry of Agriculture and Livestock and the Ministry of Water Resources were combined to from the Ministry of Agriculture and Water Resources[8]. This new institutional arrangement, under which water planning, management and development activities were placed under different ministries, made coordination and execution of any water policy very difficult. This complicated the implementation of the 1975 Water Master Plan, as well as the performance of the National Water Planning Commission[9].

By 1982, water scarcity and water pollution had become serious problems at the national level. The new administration (1982-1988) prepared a new water policy by considering these constraints. Among the other problems it addressed were flood control, conflicts between water uses and users, and low efficiency of water use in all the sectors. While the need for construction of infrastructure was acknowledged, the main objectives were appropriate use of water, maintenance of all types of infrastructure, water pollution abatement through better administration of water resources, improved social and economic efficiency, technological improvements, and human resources development.

In an effort to decentralize, the Ministry of Agriculture and Water Resources delegated activities to its offices in the states and established regional coordinating agencies to improve the integrated management of water at the river basin level. Since the new water policy emphasized the management of water resources at the regional level, it was decided that the offices of the Ministry in the states would

[8] The previous Ministry of Water Resources became one of the five offices of the newly created Ministry.

[9] Some of the main achievements of the Commission were the formulation of the overall national water policies in collaboration with the agricultural sector, formulation of the 1981 National Water Plan which emphasized regional aspects more than the 1975 version, preparation of the water plans for the states, and the basis for the water-related programmes for the 1976-1982 and 1982-1988 National Development Plans (CNA 1995).

take over the responsibilities of the river basin commissions, including further planning, management and development of water resources, and that the river basin commissions should disappear (SARH 1988). It was decided that "plans for the use of water resources, developed by the authorities at the municipal, state and federal levels, will be based on the hydrologic basin, but taking into consideration the development trends in each region." (PRI/IEPES 1982, p 152).

During the years the commissions acted as executing agencies, they made determined attempts to use water as a part of the national effort to develop the several basins and regions. However, while the activities carried out by the river basin commissions had beneficial impacts on the overall development of the country, their programmes and projects did not achieve their objectives at the regional levels. This is because increase in irrigated area would not necessarily increase agricultural production, and thus improve the quality of life of the local populations. Investment in social issues, and provision of credit and technical assistance, are equally important factors to ensure the long-term success of any agricultural development project (Barkin and King 1986).

Because infrastructure is long-lasting, the work of the commissions was appreciated mostly in those places where water, sanitation and road projects were constructed. Barkin and King (1986) have noted that the overall expenses of the Tepalcatepec Commission in the implementation of projects (Balsas Commission included), from 1947 to 1964, were 579.9 million pesos at 1960 prices. The corresponding figures for the Papaloapan Commission was 1,360.4 million pesos, for the Grijalva Commission 1030.6 million pesos, and for the Fuerte Commission 838.8 million pesos (Table 5.4).

An important factor to note for the time during which the commissions were functional, as well as at present, is that the river basin commissions, being semi-autonomous institutions, were coordinating entities for the national funds at the regional level. The commissions were active in several states, where they were able to coordinate the efforts of the several ministries to improve the social conditions of the regions. This task was specially challenging, since it depended on the political will of the parties and required considerable commitments from all the parties, which was, and continues to be, a most difficult task.

The idea of managing water at the river basin level reappeared later, and continues until the present day. However, a serious limitation that has remained until now is that the different institutions for water resources management at the river basin level continue to suffer, just like during the 1940s, from a short-term planning horizon of six years which ensures that plans, programmes and activities are not properly and adequately implemented.

Table 5.4. Expenditures of the river basin commissions, 1947-1964 (in 1960 million pesos)

Year	Papaloapan Commission	Grijalva Commission	Tepalcatepec/Balsas Commission[a]	Fuerte Commission
1947	18.6		5.7	
1948	35.8		25.8	
1949	43.5		29.3	
1950	70.9		31.8	
1951	122.3		32.5	
1952	167.4		40.9	3.8
1953	174.9	8.4	42.0	57.8
1954	137.6	13.7	35.5	154.0
1955	117.9	16.6	39.6	175.8
1956	103.1	30.5	35.5	141.4
1957	98.8	31.1	42.5	66.8
1958	110.0	37.2	35.5	38.2
1959	42.8	25.5	20.9	21.0
1960	24.3	57.9	26.6	32.9
1961	29.7	48.9	10.8	23.0
1962	20.5	239.5	21.6	24.5
1963	22.5	391.7	51.2	39.9
1964	19.8	129.6	52.2	59.7
TOTAL	1,360.4	1,030.6	579.9	838.8

Source: Ministry of Water Resources, Mexico, in Barkin and King (1986)
[a] It includes expenditures for the overall Balsas basin. Only for the Tepalcatepec basin, the expenditure for 1962 was 13.9 millions; for 1963, 13.8 millions; and for 1964, 10.4 millions.

5.4 Recent regional approach to water policies and institutions

In 1989, it was decided again that the responsibility for overall planning, management and development of water resources in the country would be under one institution, the National Water Commission (CNA by its acronym in Spanish). Initially, the CNA was under the Ministry of Agriculture and Water Resources. In 1994, it was moved to the Ministry of Environment, Natural Resources and Fisheries (Semarnap), which became the Ministry of Environment and Natural Resources from 2000 (Semarnat).

In terms of legislations, the Law on National Waters (CNA 1992; Semarnat 2004) and the Federal Law on Water Excise Taxes, set the regulatory, economic and social frameworks for water management. The Law on National Waters establishes the broad objectives for the development and implementation of the plans and the policies for water resources management. The responsibility for implementing this law has been assigned to the CNA. The Law for Ecological Balance and Environmental Protection (LGEEPA) defines the environmental regulations, and the General Health Act establishes the standards for drinking water.

Modifications to the Law on National Waters were approved on April 2004, and thus it is still too early to evaluate its impacts. One of the main changes to the Law has been the creation of River Basin Organisms, which are expected to be "autonomous units for technical, administrative and legal purposes." (Semarnat 2004, p 41). Interestingly, and inspite of the very questionable performance of the institutional arrangements at the river basin level so far (Guerrero and García-León 2003), the 2004 National Water Law states that the institutions for water management at the river basin level would continue to be the basin councils (with practically no implementing roles), that the opinions and concerns of the users would still be channelled through the general assemblies of the basin councils (which so far have not proved very useful), and that the river basin councils will be "autonomous bodies, but will be under the federal authorities."

The river basin councils, which have existed legally from 1992, but have not had perceptible impacts in terms of improving water management practices, are expected to manage water from integrated and regional perspectives, and involve water authorities at the federal, state and municipal levels, as well as the various users. For operational purposes, the river basin councils define four territorial levels: basin, sub-basin, micro-basin and aquifers, where the bodies are respectively known as councils, commissions, committees and groundwater technical committees (Marañón 2004).

The country is divided into regions and sub-regions for water management purposes. There are 13 regions, based on the hydrology of the country, and 102 sub-regions on the basis of political jurisdictions. Each sub-region includes a number of municipalities of the same state, so that regional programmes can be planned at the sub-regional level. At present, there are 314 hydrological basins, 37 hydrological regions and 13 administrative basins. Twenty-five river basin councils, out of the 26 that were planned, have already been established, but the vast majority are not yet functional for all practical purposes (Table 5.5).

By law, the river basin councils have to approve the river basin plans which, once integrated within the national water master plan, become mandatory for the federal government, and indicative for the local and the state governments and water users. Constraints, actions and resource needs are to be identified and evaluated, and unrealistic or infeasible situations are expected to be fed back into the regional planning process (Tortajada 2000).

Vast majority of the river basin councils at present are not operational: they are still in the process of organization. The councils continue to be coordinating units that can only make recommendations to the authorities and to the users. Even though the Law on National Waters (1992 and 2004) stipulates that the river basin councils can develop and implement programmes for the construction of water infrastructure which could be considered to be strategic, the fact remains that because the councils have not yet been functional, not all have say in the planning, design or operation of major water infrastructure of the country. If and when they become fully operational, it is yet to be seen as to whether the central authorities will allow the councils to use their decision-making powers on important issues, or if the traditional centralised decision-making in the country will continue for more years to come.

Table 5.5. Basin councils established as of November 2002

Name	Date established	Administrative Region
Baja California Sur	3 March, 2000	I Peninsula de Baja California
Baja California	7 December, 1999	I Peninsula de Baja California
Alto Noroeste	19 March, 1999	II Noroeste
Rios Yaqui-Matape	30 August, 2000	II Noroeste
Rio Mayo	30 August, 2000	II Noroeste
Rios Fuerte y Sinaloa	10 December, 1999	III Pacifico Norte
Rios Mocorito al Quelite	10 December, 1999	III Pacifico Norte
Rios Presicio al San Pedro	15 June, 2000	III Pacifico Norte
Rio Balsas	26 March, 1999	IV Balsas
Costa de Guerrero	29 March, 2000	V Pacifico Sur
Costa de Oaxaca	7 April, 1999	V Pacifico Sur
Rio Bravo	21 January, 1999	VI Rio Bravo
Nazas-Aguanaval	1 December, 1998	VII Cuencas Centrales del Norte
Del Altiplano	23 November, 1999	VII Cuencas Centrales del Norte
Lerma Chapala	28 January, 1993	VIII Lerma-Santiago-Pacifico
Rio Santiago	14 July, 1999	VIII Lerma-Santiago-Pacifico
Costas del Pacifico Centro	-------------	VIII Lerma-Santiago-Pacifico
Rios San Fernando-Soto La Marina	26 August, 1999	IX Golfo Norte
Rio Panuco	26 August, 1999	IX Golfo Norte
Rios Tuxpan al Jamapa	12 September, 2000	X Golfo Centro
Rio Papaloapan	16 June, 2000	X Golfo Centro
Rio Coatzacoalcos	16 June, 2000	X Golfo Centro
Costa de Chiapas	26 January, 2000	XI Frontera Sur
Grijalva-Usumacinta	11 August, 2000	XI Frontera Sur
Peninsula de Yucatan	14 December, 1999	XII Peninsula de Yucatan
Valle de Mexico	16 August, 1995	XIII Valle de Mexico

Source: Statistics on Water in Mexico (2003)

In theory, the basin councils have representatives from the federal and the state governments from the water and other sectors (Federal Power Commission, Mexican Oil Company, etc.), academia, NGOs, others and regional committees for the users of the different sectors (industrial, agricultural, drinking water providers, services, etc.). Specific commissions and committees can be established within the councils only when there are specific problems, i.e., evaluation, regional considerations, groundwater, etc. Before the amendments of the Law on National Waters, citizens and organized civil society were not part of the discussions at the river basin level (Tortajada, 2001). Even though legally this is not the case any more, only time will tell the extent of real participation of non-governmental groups, as well as their contributions.

By law, the basin councils are primarily coordinating agencies whose main role is to make recommendations to the federal government and the users on specific issues. Unfortunately, however, their establishment responds more to broader decentralization trends at the national level which still have not been achieved, rather than from the realization that water can be managed more efficiently from a regional perspective. The main problem of river basin management is not that the councils are not entitled to develop any regulations or execute any administrative or legal action, but rather that the overall operational framework to manage water resources at the basin level is still not functional in the country. If the councils are to become operational, they could play an important role as coordinating bodies. It is still not clear what are to be the operational functions of the councils, or how the councils will relate to the administrative structures of the country at the state and municipal government levels, so that they could complement, co-ordinate and support each other. At present, only one out of the 25 river basin councils that have been created is operational. In most cases, other councils do not even have staff or offices, not to mention implementable plans, financial support, or management and technical capacities (Guerrero and García-León 2003).

After more than 70 years of continuous hierarchical and top-down water management and planning practices, experiences show that the country still lacks knowledge and expertise as to how best to structure the institutional arrangements for water management at the river basin level from a decision-making viewpoint. Real participation by stakeholders has been mostly missing. Among many other constraints faced by the river basin councils are their lack of experience (both technical and managerial) as to the processes by which water policies could be formulated; reluctance of the central authorities to disseminate reliable data and information, lack of appreciation by the authorities on the importance of stakeholders' participation, and the absence of use of proper economic instruments like water pricing and demand management, and appreciation of the importance of social and environmental issues (CTMMA 2003). The main institutional challenge for the future is how best to transform the basin councils, which in practice are advisory agencies with very little real authority, into basin councils that govern, plan, organize, run, control and supervise water management at the river basin level (Guerrero and García-León 2003).

5.5 Concluding Remarks

The previous analyses show the relevance, appropriateness and impacts of the efforts made in Mexico on structuring and restructuring of its water management institutions at the river basin and regional levels. The results, unfortunately, have been mostly disappointing in terms of improving the efficiency of water resources management practices.

In the case of the river basin commissions, the overall objective was to promote economic and social growths of the appropriate regions, which were expected to reduce the prevailing inequalities among and within the regions. This was expected to be achieved through infrastructural development and settlement programmes, which, in turn, were supposed to trigger agricultural development, hydropower generation, industrial investment, employment generation and higher incomes for the local populations.

The establishment of the river basin commissions was a national effort to use water to develop the regions. However, while the programmes implemented by the commissions may have had some beneficial impacts on the overall development of the country, they seem to have basically failed to reduce regional inequalities and alleviate poverty. The decision-making failed to realize that increase in irrigated area per se was not a sufficient condition to alleviate poverty and improve the quality of life of the local people. Important issues, like investments in social services, provision of credit, technical assistance, participation of stakeholders and capacity building, were not adequately considered, even though they are absolutely essential to ensure the long-term success of any development project.

In terms of the performance of the existing institutions for river basin management so far, it has to be concluded that they have basically not achieved efficient management of water resources at the basin level. At best, they have had minor impacts in improving water management practices, which means that fundamental institutional realignments are necessary if the present basin councils are to become successful organizations to manage water regionally.

Regrettably, no comprehensive and objective evaluations have been made by the authorities concerned on the viability of such institutions, the extent to which they have fulfilled their objectives and their overall societal impacts both positive and negative. Absence of such assessments has meant that appropriate lessons have not been learnt from their failures, and/or sub-optimal performance.

While the demands for water for various uses have increased significantly in Mexico in recent decades, management practices have improved only slowly and incrementally. Consequently, the water problems of the country, in terms of quantity, quality and management, have become more serious and complex than ever before in history. The demands from different uses and users are increasing rapidly, but the technical and managerial expertise and financial resources of the concerned institutions are growing only incrementally. Unless these trends are reversed, the water situation of the country is unlikely to get better in the foreseeable future. Not surprisingly, OECD (2004) concludes that "water use in Mexico is on an unsustainable path." Furthermore, inspite of the efforts at decentralization, the real authority to plan and manage water resources continues to be vested in one single institution at the central level, which has been unable thus far to respond successfully and sensitively to the escalating needs of the sector and of the different regions of the country. Nor has it been willing to decentralize appropriate decision-making powers, investments funds and technical and managerial resources.

It is now absolutely critical to modernise the water sector of Mexico, not only in terms of institutions and laws, but also in terms of overall management practices. The water problems of the country have deteriorated due to the inadequate management, technical capacity, and the slow, and often inappropriate, responses of the sole central institution responsible for water management in the country.

Fortunately, however, the country as a whole is changing. The regions are asking for greater roles in planning, managing and decision-making not only in the area of water, but also in other sectors as well. There are some positive and encouraging indicators that some states are making good and commendable progress in planning and managing their water resources. States are realizing that water is an important resource, and its timely and proper development and management would affect the lives of the people, and their quality of life, through various pathways.

While there has been much discussion about decentralization in recent years, in practice water management has continued to be hierarchical and top-down. Unless real decentralization takes place, where institutions at regional, state and local levels would have the requisite authority, funds management capacity, and technical expertise, efficient and equitable water development and management, which could improve the social, economic and environmental conditions of the country as a whole within a reasonable timeframe, is likely to remain a distant dream.

5.6 Acknowledgements

The authors would like to thank Dr. Roberto Melville, Research Centre on Social Anthropology, CIESAS, Mexico City, and Ms. Socorro García-Sosa, Historical Archives of Water, Mexico City, for their assistance to locate historical materials for conducting this research.

5.7 References

Barkin D (1978) Regional development and reorganization of farmers. The Chontalpa as an image of the problems in the agricultural and livestock sectors in Mexico. Centro de Ecodesarrollo, Nueva Imagen, Mexico

Barkin D, T King (1986) Economic regional development. Approach by river basins in Mexico. Siglo XXI editors, 5th edition, Mexico

Bartolomé M and A Barabas (1990) Cerro de Oro dam I and II. National Council for Culture and Arts, National Institute for Indigenous People, Mexico
CNA (2003) Statistics of water in Mexico. National Water Commission, Mexico
CNA (1995) National water planning in Mexico. Experiences, results and perspectives. Ministry of Water Resources and Irrigation, Mexico
CNA (1992) Law of national waters and its regulations. National Water Commission of Mexico, Ministry of Agriculture and Water Resources, Mexico
Contreras-Moreno N (2002) Agricultural technology transfer. Assessing 60 years of experiences in Mexico, Ph D dissertation. University of Madison, Wisconsin, USA
CTMMA (2003) Water resources in Mexico. Centro del Tercer Mundo para el Manejo del Agua, AC. Porrua, Mexico
De Janvry A (ed) (1995) Strategies to mitigate rural poverty in Latin America and The Caribbean: reforms for the agricultural sector and the farmers in Mexico. International Fund for Agricultural Development and Inter-American Institute for Cooperation of Agriculture, Costa Rica
Díaz-Cisneros, H (1974) An institutional analysis of a rural development project: The case of the Puebla Project in Mexico. Land Tenure Centre, University of Wisconsin-Madison, USA
DOF (1986) Decree which abrogates the decree by which the Papaloapan River Basin Commission was created. Office of the President, 4 December 1986, Mexico
DOF (1985) Decree by which the Fuerte River Basin Commission is dissolved. Ministry of Water Resources, 8 May 1985, Mexico
DOF (1960) Decree by which the River Basin Balsas Commission is created. Office of the President, 11 November 1960, Mexico
DOF (1959) Decree by which the Fuerte River Basin Commission is created. Office of the President, 21 August 1959, Mexico
DOF (1951) Decree by which the Grijalva River Basin Commission is created. Office of the President, 29 August 1951, Mexico
DOF (1947a) Decree which creates a technical and administrative body which depends on the Ministry of Water Resources, responsible for planning, design and construction of the projects that are necessary in the basin of the Papaloapan river on 26 February 1947, 24 April 1947. Ministry of the Interior, Mexico
DOF (1947b) Decree by which the Tepalcatepec River Basin Commission is created. Office of the President, 17 July 1947, Mexico
Echeagaray L (1973) Assessment report on the Grijalva Commission, Memorandum No. 1-1024. Grijalva River Basin Commission, 11 December 1973, Mexico
Estrada V (2003) Management of river basins in Mexico. The Papaloapan Commission, 1947-1985, Bachelor Thesis. UNAM, Mexico
Grijalva Commission (1975) Report on the activities between 1 November 1974 and 31 October 1975. Fifth Report of the state government, Mexico
Guerrero V and F García-León (2003) Proposal for the decentralization of water management in Mexico by means of basin councils. In: Water Policies and Institutions in Latin America, Tortajada C, Braga BPF, García LE and Biswas AK (eds) Water Resources Management Series. Oxford University Press, Delhi, pp 144-159
Herrera-Toledo C (1997) National water master planning in Mexico. In: National Water Master Plans for Developing Countries, Biswas AK, Herrera-Toledo C, Garduño-Velasco H and Tortajada-Quiroz C (eds) Water Resources Management Series. Oxford University Press, Delhi, pp 6-53

McMahon D (1989) Anthropology of a dam, the mazatecos indigenous people and the Papaloapan project. National Council for Culture and Arts, National Institute for Indigenous People, Mexico

Marañón B (2004) Social participation in the management of groundwater: rhetoric and reality. In: Towards an integral management of water resources in Mexico: challenges and alternatives, Tortajada C, Guerrero V and Sandoval R (eds). Porrua, Mexico, pp 231-256

Murillo D (2003) Planning for the sustainable development of the Papaloapan river basin. Mexican Institute for Water Technology, Mexico, http:://water.nml.uib.no/project/?id=144

OECD (2004) OECD Economic Surveys. Mexico, 2002-2003. Organisation for Economic Co-operation and Development, Paris

Orive A (1970) The irrigation in Mexico. Grijalbo, Mexico

Poleman T (1964) The Papaloapan project, Agricultural Development in the Mexican Tropics. Stanford University Press, California

PRI/IEPES (1982) Working groups for the development of the government plan 1982-1988 on water issues, preliminary document. Institute for Political, Economic and Social Studies, Mexico

Redcliff, M., 1982, Development policymaking in Mexico: the Mexican Food System. University of London, Institute of Latin American studies, London

SARH (1990) Proceedings of the Papaloapan River Basin Commission, 1947-1988, Volume I and II. Ministry of Agriculture and Water Resources, Mexico

SARH (1988) Water and society. Ministry of Agriculture and Water Resources, Mexico

SARH (1986) Proceedings of the Fuerte River Basin Commission, 1951-1986. Ministry of Agriculture and Water Resources, Mexico

Schumacher A (1981) Agricultural development and rural development: A Mexican dilemma. University of Maryland, Maryland

Semarnat (2004) Decree by which the National Water Law is modified. Ministry of Environment and Natural Resources, DOF 29 April 2004, Mexico

SRH (no date) Grijalva basin. Ministry of Water Resources, Mexico

SRH (1975) National Water Plan. Ministry of Water Resources, Mexico

SRH (1972) Papaloapan Commission 1947-1972. Ministry of Water Resources, Mexico

SRH (1961) Proceedings for the activities carried out by the Tepalcatepec Commission, 1947-1961. Ministry of Water Resources, Mexico

SRH (1958) The economy of the Papaloapan Commission. Ministry of Water Resources, Mexico

SRH (1952) Report of activities 1 September 1951 to 31 August 1952. Ministry of Water Resources, Mexico

Tortajada C (2001) Institutions for integrated river basin management in Latin America. Water Resources Development, 17:289-301

Tortajada C (2000) River basins: Institutional framework and management options for Latin America. World Commission on Dams, Thematic Reviews, River Basin Institutional Framework and Management Options V3, Cape Town

Wester P, R Melville and S Ramos-Osorio (2001) International arrangements for water management in the Lerma-Chapala basin. In: Hansen A and van Afferden M (eds) The Lerma-Chapala watershed: evaluation and management. Kluwer Academic/Plenum Publishers, New York, pp 343-369

Wionczik MS (1992) Contributions of the water policies between 1925-1970 to the present Mexican crisis in the agricultural sector. Comercio Exterior, 32 (1): 394-409

Water Institutions in the Middle East

Peter Beaumont

6.1. Introduction

Water institutions probably originated around 10,000 years ago with the advent of irrigated agriculture in the Middle East. Their role was to ensure efficient allocation of water in a particular river basin or locality. At first they would be small and simple to administer, but their complexity increased as irrigation systems developed. When a particular locality was overrun by a new political group it would seem that the existing water institutions were adopted with few changes. After all they worked and so there was little incentive for changes to be instituted. With the advent of Islam the same process appears to have taken place and water institutions were incorporated within a broader religious framework. Basic structures, however, appear to have been little changed. In the twentieth century, with the advent of nation states water planning bureaucracies based in the capital cities have been superimposed over existing water institutions and it is only in the last few decades that traditional water institutions have begun to be replaced by more modern methodologies.

6.2 The Beginnings of Water Institutions

Agricultural activity in the Middle East began around 12,000 years ago. Before this time human groups exploited a wide spectrum of resources and so located themselves at places that would provide access to a range of different environments. Once agriculture was developed, however, the chief locational factors for human activity were deep soils and adequate water resources. The earliest agriculture probably took place at sites where rainfall provided the necessary water. However, these early agricultural societies quickly realized that higher yields could be produced and more reliable crop production achieved by the use of irrigation. Just when this first use of irrigation took place is unknown but investigations at the ancient site of Jericho suggest that irrigation was being practised as long as 10,000 years ago (Kenyon 1969).

The significance of irrigation is enormous as it implies that some form of organization within society must be taking place to ensure that the watering of the crops is successfully achieved. This is particularly the case if a number of farmers are obtaining their water supplies from a single water source. It is, therefore, possible to suggest that the earliest water institutions came into being 10,000 years ago in areas such as the Dead Sea lowlands. These earliest water institutions, like their counterparts at the present day, faced the critical problem of how to allocate the available water amongst the competing claims of the local farmers. It is not known how this was actually resolved, but evidence from traditional water division systems in Iran would suggest that it might have been on a time-share basis (Lambton 1953). This has the advantage that when river and stream levels are high, all people get more water in a particular time period. Similarly with low stream flows everyone suffers equally as water volumes decline per unit of time.

It is important to realize that water institutions must have existed within human societies from the earliest days of irrigation. However, while irrigated agriculture remained a constant in a region the societies and their political, economic and religious beliefs would have been subjected to considerable change. In many cases indigenous societies would have been overrun by belligerent neighbours. What is interesting here is to speculate on the extent to which a conquering ruling class would attempt to change the water institutions it inherited. If these institutions appeared to be working successfully it would seem likely that the newcomers would incorporate them into their own system of governing the conquered peoples. This would be highly likely if the invading peoples came from a region where irrigation was not common or if the invading society was urban based. If this practice of the acceptance of local water institutions is correct it would suggest that many of these organizations in any region might be able to trace their origins back over long time periods. In other words it would not really be correct to talk of Archaemenid, Greek, Roman or even Islamic water institutions as each succeeding political entity merely inherited and developed what were already existing organizations.

Common sense would suggest that each major region would develop water institutions that were able to cope with the range of local problems that the farmers had to deal with. The larger the water planning region the more complex an institution would have to be to handle the likely problems which arose. Interesting situations must have developed in the larger river basins when different water users along the river might have begun to compete for the available water resources at a particularly critical period in the calendar of plant growth. In such cases agreements as to how much water each region should have access to may have been reached as the result of negotiations between the interested parties or a solution may have been imposed on the parties concerned by a dominant ruler.

Traditional water institutions, many of which may have dated back to pre-Islamic times, continued to operate in parts of the Middle East until at least the time of the Second World War. For example, the traditional allocation of irrigation water on the River Zayandeh in the Isfahan oasis was based on a code known as the "Tumar" (Lambton 1953). This code had been developed through local custom over the years and it survived well into the twentieth century. The key feature of the "Tumar" was that between late November and the end of May anyone could abstract as much water from the river as they wished. During the rest of the year when water was scarce strict controls governed water use. The water of the river was divided into 33 primary shares and then these were allocated to the main districts in the Isfahan oasis (Beaumont 1993:286). The 33 major shares were subdivided into 276 secondary shares associated with the major irrigation canals and into 3,105 tertiary shares at the village level. Given the complexity of the water rights, the successful operation of the system depended totally on the skills and honesty of the officials administering them. The maintenance of the water distribution system was a communal responsibility with the farmers maintaining the particular canal from which they received their water.

The way in which the system operated meant that the upstream districts of the oasis, Lenjan, Alenjan, Marbine and Jey, were the most privileged in terms of the water they received. In contrast, the downstream districts of Isfahan, Kararadj, Baraan and Rudashtin were only allocated water at the beginning and end of the growing season. This did cause problems for the downstream districts, especially in years of water shortage, but at the same time it did ensure the continued prosperity of the upstream districts.

6.3 Islam and Water

As many of the people of the Middle East follow the Islamic faith it is useful to examine the impact that the teachings of Islam have had on water institutions and water management. Over the last few years an increasing number of papers have appeared dealing with the Islamic view of water and environmental management (Faruqui et al. 2001). What is obvious from this literature is that the Koran and the hadiths, the basis of Islamic law, have relatively little to say directly about water and water management. For example, the word "water" occurs only sixty-three times and "river" or "rivers" fifty-two times in the Koran (Amery 2001:40). However, perhaps the most quoted sentence from the Koran is "*And We created from Water every living thing*". This clearly indicates that water is regarded as a vital life force and a unifying factor in Islam. Equally the verse "*And Allah has sent down the water from the sky and therewith gives life to the earth after its death*" stresses once again the life giving quality of water.

However, the relative lack of specific references to water means that many statements of a more general nature have had to be interpreted within the context of water and its management. A very important concept in Islam is that of stewardship of the earth and its resources and the need to think about the requirements of future generations. "*Act in your life as though you are living forever and act for the Hereafter as if you are dying tomorrow*" (Izzi Dien 1990:194). From this Islamic scholars would claim that the modern concept of "sustainable development" fits in well with Islamic ideals. Equally the statement "*Make not mischief on the earth*" is regarded by certain scholars as encompassing the need to prevent environmental exploitation, degradation and pollution (Amery 2001:41).

The lack of specificity in many verses of the Koran and in the sayings of the Prophet Muhammud does have its advantages. It means that certain statements can be re-interpreted in the light of modern developments. For example, it had originally been thought that Islamic teaching meant that wastewaters could not be used for a range of purposes. However, in 1978 the Council of Ulamah in Saudi Arabia issued a Fatwa concerning the conversion of wastewater into pure water (Abderrahman 2000:516). This stated that: *"impure wastewater can be considered as pure water and similar to the original pure water, if its treatment, using advanced technical procedures, is capable to remove its impurities with regard to taste, colour, and smell, as witnessed by honest, specialized and knowledgeable experts. Then it can be used to remove body impurities and for purifying, even for drinking. If there are negative impacts from its direct use on the human health, then it is better to avoid its use, not because it is impure but to avoid harming the human beings"* (CLIS 1978:40-41). This dynamic nature of Islamic law does mean that individual concepts can be re-interpreted in the light of increasing scientific knowledge.

In certain cases, however, decisions about specific water matters have to be reached on more pragmatic grounds. With regard to Islamic law and custom, priority with regard to access to water is given first to humans for drinking purposes, secondly to the watering of animals, and thirdly for agricultural purposes. Other, what might be described as more modern uses, was not considered. To deal with such situations new approaches had to be devised. "*In Saudi Arabia, industrial and recreational uses (of water) come fourth and fifth respectively. The order of the last two purposes was ranked according to the application of Islamic customs in the country and through reasoning rather than from strict doctrine*" (Abderrahman 2001:69).

Islamic law includes the principle of "no injury" or "no harm"(Amery 2001:45). This idea is widely applied to mean that the actions of one human being should not produce a detrimental effect on another one. It is easy to see that this principle can be applied in a water resource context in terms of the avoidance of waste and the prevention of pollution.

Overall it is clear that the teachings of Islam, with their emphasis on respect for the environment, the avoidance of waste, and causing "no harm" to others fit in well with ideas on sustainable development. It certainly includes all the concepts and ideas which are needed to construct a water management policy that is in touch with the needs of the modern world. Equally important is the fact that the guardians of Islam, the leading scholars of the day, have shown that they are willing to re-interpret the meaning of the Quran and the sayings of the Prophet Muhammud to meet the needs of modern societies.

From an economic point of view many Muslim scholars have divided water resources into three categories (Kadouri et al. 2001:89). These are private goods, restricted public goods, and public goods. Water which is stored in a container, such as a cistern or reservoir, or in specially constructed infrastructure systems, such as irrigation canals or wells, is regarded as a private good belonging to the owner of the facilities. This owner can do what he wishes with the water, including selling it to another person. Interestingly, water which has been collected and treated in a wastewater plant can also be sold as the organization dealing with this water has obtained ownership rights through the investment in a collection system and plant to treat the water.

When a water body such as a stream or small lake is situated on lands that are privately owned it is considered to be restricted public goods. In this case the landowner does not "own" the water, but merely enjoys special rights and privileges compared with other users. However, anybody can use the water for drinking purposes and to satisfy basic needs. Agricultural or industrial use of the water cannot take place, however, without the permission of the landowner.

Water found in large rivers, lakes and major aquifers are considered a public good. Anyone can use this water for drinking, agricultural and industrial purposes provided that it does not cause environmental degradation or public harm. If this water is transported via canals, pipes or containers it becomes the property of the person who has invested in the infrastructure.

In modern times it would seem that the nation state has become the de facto controller of public goods such as water. The logic here being that the nation state has taken over the role of the Imam or ruler. This has meant that traditional water uses have been allowed to continue by governments provided that they caused no serious environmental damage nor harm to society in general.

Where water is being sold, Islamic custom decrees that it is the market mechanism that sets the price. In such cases governments are not expected to intervene unless there is evidence of price fixing and manipulation of market conditions by unscrupulous merchants (Kadouri et al. 2001:90). An important tool in modern water resource management is the management of demand through the use of the pricing mechanism. As this is applied in a modern context its aims are usually to cut down the wastage of water and at the same time to ensure social equity. As such its aims fit in well with Islamic ideals, particularly if it provides the poor with better access to water.

6.4 Modern Water Institutions

With the advent of nation states in the twentieth century the responsibility for water resource management has been transferred from the local or regional level to the national level. A feature of all the modern Middle Eastern countries is a large state bureaucracy, with power being exercised through strong central control from the capital city. This means that the ways in which water is managed at the present time in the Middle East reveals remarkable similarities from one country to another. The names of the agencies may vary from country to country, but the objectives being pursued remain basically the same. In a sense, this is not surprising, as the problems being faced by these countries are similar. The only difference is the particular stage of development that has been reached.

In certain countries the development of what might be termed modern water institutions dates back to the nineteenth century. For example, in Egypt, the Nizarah of Public Works was established in 1857 and this included the Agriculture Department, the Sewage Department and the Irrigation Department (Ministry of Water Resources and Irrigation, Egypt, 2002). The name Nizarah was changed to Ministry in 1914 to form the Ministry of Public Works. From the 1960s onwards a number of name changes have taken place with regard to the major institutions dealing with water. In 1964 a new ministry was formed and given the title Ministry of Irrigation. In the late 1970s (1977) responsibilities for land reclamation were given to the Ministry of Irrigation and its title was changed to Ministry of Irrigation and Land Reclamation. Further name changes occurred in 1987 when it became known as the Ministry of Public Works and Water Resources and in 1999 when the new name of Ministry of Water Resources and Irrigation was adopted.

All of the countries of the Middle East have now reached the stage when all the easily available water resources have been developed. Extra water is still available in some countries but its development will be costly. Therefore, the key task facing all nations is how to allocate the available water resources to produce maximum economic productivity and social justice for their people. Until recently, most of the countries of the region have been using up to 90 percent of their water resources on irrigation to produce crops with a relatively low economic value (Beaumont, 2000a). With the growth of the urban systems in the Middle East, most of the wealth-creating capacity is now located in these regions. Yet many of these towns and cities suffer from severe water shortages and even water rationing which obviously hinder the growth of wealth creation. The great challenge for the water resource institutions of the various countries is to develop policies that will ensure the continued economic growth of theses cities by ensuring the delivery of adequate supplies of water. In most cases this can only be achieved by the cutback in water which is used for irrigation. As yet most of the water institutions have not come up with clear policies for how this is to be achieved.

6.5 National Case Studies

To illustrate the ways in which the countries of the region have approached their water management issues a series of national case studies have been included. These deal with Iran, Israel, Jordan, Saudi Arabia and Turkey.

6.5.1 Iran

Iran is a country with a wide variety of different environments and a long history of human settlement. Much of the country experiences water shortages during the summer months and so it seems likely that water institutions of one form or another must have existed here for thousands of years.

From the formation of the modern state under Reza Shah in the 1920s until the Islamic Revolution of 1979 the major influences on water use in Iran were ancient customary law, Islamic law, as interpreted in the Shi'a tradition, and the Civil Code. In effect the Civil Code followed closely the rulings and concepts of Islamic law. People who constructed qanats and irrigation canals became the owners of the water, and this water would then be subject to the laws of inheritance. With larger bodies, such as rivers and lakes, water ownership was held in common in the Islamic tradition, but in general the upstream landowner held precedence over those further downstream. Similarly, the doctrine of prior appropriative rights was observed with an established water user taking priority over a newcomer. This idea was confirmed in the Civil Code (McLachlan 1988:70).

The Civil Code also paid regard to the maintenance and repair of water supply facilities such as irrigation canals and qanats. When disputes over maintenance arose the courts could pass judgement on them and co-owners who were not willing to contribute to the maintenance of a particular water facility could be fined (McLachlan 1988:71). Unfortunately, such judgements were not always carried through which meant that many qanats and irrigation canals fell into disrepair.

McLachlan claims that although Islamic law has theoretically governed all water law under both the monarchy and at least the early part of the Islamic Republic of Iran the reality was somewhat different (McLachlan 1988:71). He pointed out that the actual control of the irrigation systems was the responsibility of local officials who occupied traditional roles and used a series of rules and measurements for the management of water supplies that were probably hundreds of years old. Many of these customary practices, it is believed, even dated back to pre-Islamic times.

Lambton also believes that traditional customary practices were accepted following Islamic control of a region, partly because Islamic civilisation was mainly urban-based and less concerned with the details of agricultural activities (Lambton 1981:283-288). The importance of customary law was recognized by the state in the Civil Code adopted by Reza Shah (McLachlan 1988:72). In effect, customary laws were accepted as applying in all cases where no new legislative provision had been made.

During the time of Reza Shah innovations were made with regard to water law. In 1930, a qanat law was enacted to facilitate the construction and protection of qanats. This was followed in 1937 by a law to develop village land and water resources under the guidance of regional councils (McLachlan 1988:72). Unfortunately, the power of the landlords and the inefficiency of the state bureaucracy meant that these acts achieved very little. A more important action was the establishment of the Independent Irrigation Institute in 1943 with the role of managing the conservation and development of water resources at a national level.

Until the 1950s a large proportion of the water supply over much of the central plateau of Iran had been delivered by qanat systems (Beaumont 1971; Beaumont et al. 1989). The allocation of water from the qanats followed complex distribution patterns that had evolved over the years. On the Iranian plateau where irrigation by qanats is important, a characteristic pattern of water management can be observed. Within the larger villages or groups of villages the person in charge of water distribution was known as the mirab, with a number of other officials working under him. Water officials lived in the villages and their positions were usually permanent ones. They could be paid in cash, in water shares or even in crops by the water users. The mirab kept a record of the water shares of each farmer in his notebook. This notebook would also record any trading, buying or selling of water shares (Bonine 1977:21). The traditional way of timing the water shares was by a water clock, or sundial, but since the 1950s these have been replaced by first spring-driven clocks and later by quartz watches. The cycles for the individual qanats could vary with many having cycles of between 8 and 13 days. In larger villages and towns there is considerable trading and selling of water shares. All of these have to be recorded by the mirab. In order to keep the qanats flowing at all times a team of men was required to clear away sediment and make good any tunnel collapses. Such men were usually paid by receiving a share of the water from the qanat or qanats which they looked after. This water could then be used on their land or sold to another user.

From the 1950s onwards the use of groundwater increased in Iran through the introduction of pumped wells. These were relatively cheap to sink and to operate. The main problem with them was that they often caused the water-table to decline and as a result the discharge of many qanats declined or ceased completely (Beaumont 1968). The government recognized this problem in the 1960s and brought in legislation which forced people to apply for a permit before a new well could be sunk.

The nationalization of water resources in Iran took place in 1968 as part of the Shah's "White Revolution" or land reform programme. In the Water Nationalization Act the responsibility for managing water resources in Iran was given to the Ministry of Water and Power, while at the regional level a series of boards were established to collect data, to restrict water consumption and to charge water rates. The objectives of the Water Nationalization Act are clearly stated in Article 1 (Echo of Iran 1971:375).

Article 1: "All water flowing in rivers, natural streams, valleys and other natural water courses, whether surface or subterranean, and also all flood waters, sewage waters, and drainage waters, as well as waters of natural lakes, swamps, lagoons and ponds, springs, mineral waters and subterranean water deposits, are considered natural wealth and belong to the public. The Ministry of Water and Power is assigned the task of protecting and utilizing this national wealth, and of setting up and administering installations for the development of water resources."

This water nationalization legislation gave the Ministry of Water and Power much greater powers than had been enjoyed by the Independent Irrigation Institute. In particular, the Ministry was able to charge for water that was used below the major dams and so ensure that wasteful use was minimized. This was in recognition of the fact that state funding of major water infrastructure projects would only make sense if some of the capital could be recovered by a water charging policy. Prior to this, the managers of the major dams had been reluctant to release water for agriculture as higher financial returns could be obtained by using the water for power production or urban water supply. This pricing policy complied with Islamic teachings in that the water became "private property" as it had been captured by the construction of special facilities such as dams.

During the 1960s the Iranian government also embarked on a major development project of Khuzestan using water from a series of large dams which had been constructed in the region. This development was carried out under the guidance of the Khuzestan Water and Power Authority that was established in 1959. A number of agribusinesses were established associated with the dams, often with foreign capital, but the results of their operation were disappointing owing to inadequate land preparation and over-watering problems.

Towards the end of the Shah's reign in the late 1970s the Iranian government drew up a National Water Plan for the first time. This was an ambitious and forward-looking document which attempted to make a comprehensive assessment of the water problems facing Iran. The objectives were:

1. Formulate a National Water Plan concept which emphasises long-term water resource management
2. Determine, to the extent possible, the quantities, qualities and sources of water resources available in Iran
3. Determine the demands upon the water resources by present and future population and economic activities
4. Prepare a comprehensive overview of the impact that present governmental plans, goals, objectives and policies may have upon the availability of the nation's water resources if they were to be fully developed
5. Recommend an institutional framework and organizational structure which will facilitate effective water resource planning on a national scale
6. Recommend actions and/or programmes to alleviate or solve identified water and related land resource conflicts and problems

The main organization dealing with water management in the National Water Plan was to be the Bureau of Water Planning located in the Ministry of Energy

(Bureau of Water Planning and Development and Resources Corporation 1978:37). Its chief role was to encourage the optimum use of water resources in conjunction with public agencies and private enterprises. Beneath the Bureau of Water Planning were to be a series of regional water authorities with a wide range of responsibilities. These authorities would implement the plans established by the Bureau of Water Planning and be responsible to it. Overall, the National Water Plan for Iran revealed a genuine desire to develop a more sustainable management of the water resources of the nation. Unfortunately, changing political conditions and the fall of the Shah meant that the implementation of these reforms had to be postponed.

The new Islamic Republic of Iran that came to power in 1979 closely followed the teachings of Islam in the ways in which the government operated. Nevertheless, many of the water management policies implemented by the previous regime were followed. In the new constitution of the Islamic Republic of Iran Article 45 (Public Wealth) states: *"Public wealth and property, such as uncultivated or abandoned land, mineral deposits, seas, lakes, rivers and other public waterways, mountains, valleys, forests, marshlands, natural forests, unenclosed pastures, legacies without heirs, property of undetermined ownership, and public property recovered from usurpers, shall be at the disposal of the Islamic government for it to utilize in accordance with the public interest. Law will specify detailed procedures for the utilisation of each of the foregoing items"*. This clearly states that the new government has the right to decide what should be done with public property and in this sense continues the ideas set out in the National Water Plan of the previous government. In 1982 the new Islamic Republic of Iran revised the water nationalization law and renamed it as the "Just Distribution of Water Law". This law remained similar to the earlier one, but payments for interest, which had been permitted in the earlier law, were not included.

During the 1980s the government of Iran was faced with increasing pressures on the available water resources as population numbers grew rapidly. It decided that new water charging policies had to be introduced which minimised wasteful uses of water.

As a consequence, a new water charging policy for agriculture was devised in 1990 based on the type of canal network in use. With modern networks, usually primary and secondary canals from major dam schemes, the charge was 3 percent of the revenue of the crops planted (Sadr 2001:110). With traditional canals it fell to 1 percent, and when a combination of modern and traditional was used the charge was 2 percent. At the same time, a supervision charge for the use of groundwater was introduced. For example, with wheat, a charge of 0.25 percent of the price obtained for the crop is made (Sadr 2001:112).

1990 also saw the passing of an act for the establishment of municipal water and sewage companies in Iran. What is interesting is that both the private and public sectors were free to invest in and to manage these plants, although the companies have to operate in accordance with the Iranian Trade Law. The idea is that these plants should be largely self-financing with charges for water reflecting operational and depreciation costs. The water charging policy that is used is based on a block system. Households with a consumption of less than 5 m^3 per month levied no charge so that poor families have access to sufficient water to meet their basic needs. Thereafter, charges increase for each successive block of around 10 m^3. From 1998, charges for commercial and industrial water use were set at a higher value than for residential use (Sadr 2001:110).

In Iran at the present day a complex system of water management is in existence with elements drawn from ancient customary law and the majority based on tenets of Islamic law. The interpretation of Islamic law has, however, been sufficiently broad to permit the continued private ownership of water to coexist alongside the general concept of water as a public good. The idea of selling shares in water has been long established in Iran and this practice continues at the present day. With regard to infrastructure provision for water management a pragmatic solution appears to have been reached with large projects for the provision and preservation of water carried out by the public sector, while transfer and distribution of water is undertaken by the private sector. The main organization charged with dealing with water resources in Iran at the present day is the Water Resources Management Organization. This institution is responsible for all aspects of water policy in Iran and forms part of the Ministry of Energy.

The main challenge for the Water Resources Management Organization in future will be how to supply sufficient water for the urban centres in which most of the population of the country resides. To achieve, this water will have to be reallocated from agricultural to urban use, but how this will be achieved is not yet certain. The sale of water to the highest bidder looks as if it might be the most efficient method to achieve this objective.

6.5.2 Saudi Arabia

Saudi Arabia follows Muslim law or "Shari'a" in all aspects of its life and government (Abderrahman 2000:513). All laws and regulations dealing with water that have been introduced follow the principles of Islamic law. From the Shari'a it is considered that all Muslims have a common entitlement to water. In the case of Saudi Arabia, this is taken to imply that the "state" represented by a Muslim government has a duty to supply water to its citizens.

In the period since the Second World War, the Government of Saudi Arabia has devised comprehensive water policies and established government agencies to supply the needs of its citizens. These policies and agencies have often taken over the roles of earlier institutions operating at the local level. The Ministry of Agriculture and Water was set up in 1953 as the prime organization dealing with water in the Kingdom. During the next decade, when it was realized that desalination would play a major role in future water policy, the Saline Water Conversion Corporation was established as an agency under the Ministry of Agriculture and Water in 1965. In 1974 the Saline Water Conversion Corporation was made an independent corporation within the Ministry of Agriculture with responsibility for the construction and operation of desalination plants for drinking water production. The distribution of drinking water and the collection and treatment of wastewaters in urban centres was carried out by Water and Wastewater Authorities which were independent agencies under the Ministry of Rural and Municipal Affairs.

In recent decades the Saudi government has faced considerable water problems that have placed severe strains on the available resources. In part, this has been the result of rapid population growth. Since 1980, water demand has increased from 2,362 million m^3 to 18,496 million m^3 in 1997. This represents a growth of more than seven-fold in less than two decades. To help meet this demand, more than 30 major desalination plants have been constructed along the Red Sea and Gulf coasts so that by 2000 the installed capacity was 1,050 million m^3 per annum (Beaumont 1977; Abderrahman 2000:514).

The main problem for Saudi Arabia in future will be the provision of domestic water. Already, growth in this sector has been considerable. Demand in 1980 was only 446 million m^3 per annum, but by 1997 it had risen to 1,563 million m^3 and it is projected to reach 2,800 million m^3 in 2010 (Abderrahman 2000:515). From 1994 Saudi Arabia has employed a water-use tariff which has increased steeply with greater water use in an attempt to curb demand. For the first 100 m^3, the charge is 0.04 US$/$m^3$. For the second 100 m^3, it rises to 0.27 US$/$m^3$. For the third 100 m^3, it is 0.53 US$/$m^3$ and for the fourth 100 m^3, it is 1.07 US$/$m^3$ (Abderrahman 2000:515).

A remarkable feature of the main water institution in Saudi Arabia, the Ministry of Agriculture and Water, has been the way in which it has changed its policy on irrigation development over the last twenty years or so. From 1979 onwards the government supported a programme of large-scale agricultural well drilling and modernization of the irrigation infrastructure. It is interesting to note that in 1980 the government introduced a decree which meant that any new wells could only be sunk following the issue of a permit from the Ministry of Agriculture and Water (Abderrahman, 2000:516). In other words, the Ministry already possessed the ability to control the utilization of the country's aquifers and to ensure that a policy of sustainable development was followed. Instead, for strategic reasons, it chose a policy designed to achieve food self-sufficiency.

As a consequence of what might be described as an exploitative environmental policy, the expansion of agriculture, largely for wheat production, meant that water use for irrigation expanded from around 1,850 million m^3 in 1980 to 29,826 million m^3 in 1992 (Abderrahman 2000:515). This more-than 15 times increase was the result almost entirely of the abstraction of fossil water from the great aquifer systems that underlie the Kingdom. Fortunately, the government soon came to realize that its wheat policy, including huge financial subsidies to farmers, was unsustainable owing to the detrimental effect it was having on the groundwater resources. In 1993 the government reduced its subsidies for wheat production in an attempt to match wheat production to domestic needs only. By 1997, irrigation water use had dropped by almost a half to 16,406 million m^3 and by 2010 it is hoped that it will have fallen further to 14,700 million m^3 (Abderrahman 2000:515).

From the late 1990s, the Ministry of Agriculture and Water has begun to pursue more sustainable water policies. Wastewaters are now being collected and treated and used for irrigating date palms and forage crops. This practice is expected to increase in the future. Equally, the urban populations are being encouraged to use less water by the block pricing policies mentioned above. However, there is still a long way to go, as even with this pricing system, the revenues obtained from water supply only cover about one-fifth of the actual water costs.

6.5.3 Israel

Of all the countries in the Middle East, Israel is the only one that does not have the majority of its population following the Muslim faith. It is also one of the youngest countries in the Middle East and, as a result few, of its water institutions reveal the historical continuity that is so characteristic of most other countries in the region. It is, therefore, instructive to analyse the water institutions that came into being as this young country has evolved.

When Israel was established in 1948, the country was still at war with its neighbours and so strategic considerations played an important part in the early development of the country. One of the ways in which this showed itself was a government policy of encouraging as much rural settlement as possible, particularly in the border regions and the Negev. The belief here was that this would improve security and reduce the risks of attack from Palestinians who had been displaced form their lands. From a water point of view, it meant that Israel became committed from an early stage to the widespread development of irrigated agriculture.

The most important legislation dealing with water in Israel is the 1959 Water Law. In this Law, Israel, in effect, nationalized its water resources. The first clause of Chapter One states: *"The water resources of the State are public property; they are subject to the control of the State and are destined for the requirement of its inhabitants and for the development of the country"*. This meant that from this time onwards the possession of land rights did not confer any rights to water sources on that land. With regard to the rights of individuals Clause 3 says: *"Every person is entitled to receive and use water, subject to the provisions of this Law"*. This is further qualified by Clause 4 which goes on to say: *"A person's rights to receive water from a water resource is valid so long as the receipt of that water resource does not lead to the salination and depletion thereof"*. A particularly interesting clause is Clause 6 as it links water use to a specific purpose: *"Every right to water is linked to one of the water purposes enumerated hereunder; the right to water ceases upon the cessation of the purpose. The purposes are: (1) domestic purposes; (2) agriculture; (3) industry; (4) handicraft, commerce and services; (5) public services"*. The efficient use of water is ensured by Clause 9, in Chapter 2: *"A person shall: (1) deal efficiently and sparingly with water coming into his control; (2) keep any water installations under his control in proper condition so as to prevent the waste of water; (3) refrain from obstructing or depleting any water source; (4) refrain from the obstruction or depletion of a water resource from which he produces water"*.

The Minister of Agriculture was given responsibility for ensuring the Water Law was implemented, with the assistance of a Water Commissioner and the Water Council. Given the dominant role of the Minister of Agriculture in water policy planning in Israel, it is not surprising that water for irrigation has always received a high priority in water allocation decisions. When the Ministry of the Environment was set up, responsibilities for protecting water quality and preventing water pollution were transferred to this new Ministry. The Minister of Agriculture remained responsible for water use and can prescribe norms for the quantity, quality, price, conditions of supply, and use of water (Ministry of Foreign Affairs, Israel, 2002). The Minister of Agriculture was also responsible for the efficient use of water and has the power to introduce rationing, when necessary.

The Water Commissioner, who is appointed by the Government, is responsible for the enforcement of the Water Law and Water Regulations and for ensuring water quality. In effect, the Water Commissioner is in charge of managing the Israeli water system. As such, the Water Commissioner does play a critical role in the implementation of water policy. However, it is interesting to note that the various Water Commissioners over the years have often had very different views on the directions in which water policy in Israel should be heading.

The Water Council is chaired by the Minister of Agriculture, and has the Water Commissioner as the Deputy Chairman. Its main duties are to advise the Minister of Agriculture on matters of policy and prevention of pollution and to approve regulations concerning water quality, price, conditions of supply, and rationing. Over the years, the majority of the members of the Board have had connections with the agricultural sector; a fact which must obviously have influenced the decisions taken by the Board. There is also a Tribunal of Water Affairs, established by the Ministry of Justice, which may issue fines or prison sentences to those who contravene the provisions of the Water Law or the Drainage and Flood Control Law.

The implementation of water policy in Israel is carried out largely by Mekorot, a non-profit, public corporation that was established in 1937, before the creation of the state of Israel. Mekorot is the most powerful water institution in Israel and effectively acts as a National Water Authority. Mekorot is responsible for the construction, operation and maintenance of the water infrastructure in Israel and for licensing water use in the different sectors of the economy. It delivers at least two-thirds of the water supplied in Israel, and since 1967, has controlled all water resources in the occupied territories. With near-monopoly powers, Mekorot has become a highly centralized and bureaucratic organization and, like all such organizations, has become resistant to change. Tahal (Water Planning for Israel Ltd.) is a non-profit government corporation that was established in 1952. It is the main water planning and research organization in Israel and acts as an advisor to Mekorot with regard to major water projects.

Water is also supplied by regional water associations that were originally set up to supply irrigation water. These can operate in conjunction with Mekorot or independently. Water associations are often linked to the political parties drawing their support from the Kibbutz and Moshav movements. Such movements strongly support water being used for irrigation purposes, as these settlements utilize around 85 percent of all irrigation water (Soffer 2002:10). There also exists within Israel an Equalisation Fund, which, in conjunction with the Water Commissioner, regulates price differences in the various regions of the nation. In effect, it attempts to make water prices similar, irrespective of the true cost of water supply.

It is important to realize that many other Ministries and special interest groups in Israel also have an influence on water policy. An important player is the Ministry of Finance that has overall responsibility for all spending in the country. For many years, this Ministry has held the view that the raising of water prices for farmers is the best way to deal with the water crisis (Soffer 2002:4). The Ministry of Health becomes involved in the water industry through the quality standards required for drinking water and in terms of the standards to be attained for wastewaters to be used for irrigation and industrial uses. The Ministry of the Interior is responsible for controlling the finances of the local authorities and so this, in turn, has an impact on water supply and sewage treatment carried out by them.

The nature of the water institutions in Israel and how the various responsibilities have been allocated to the various Ministries goes a long way to explain the priority that was given to irrigation water in the first few decades of the country's existence. This did not cause any problem up to the late 1960s as, up to this time, the volumes of water available were greater than water demands. However, over the last 30 years, Israel has faced greater pressures on its water resources that have, in effect, been self-inflicted. Until the last few years, Israeli agriculture has continued to enjoy a favoured position with regard to water resource use. Indeed, it was only in the 1990s that the decision makers in Israel began to realize that the use of high quality water for irrigation was no longer sustainable. They quickly realized that if Israel was to continue to prosper, irrigation water would have to be reallocated to urban/industrial uses.

This change of policy was, however, very difficult to implement as long as the Minister of Agriculture enjoyed such an important role in water policy decision-making. It is equally important to realize that it has always been in the political interests of Israel to exaggerate its water problems as this provided an excuse for denying water to the Palestinian citizens of the territories which Israel had invaded and occupied in 1967.

Over the last few years, perceptions about water use have begun to change in Israel, and growing numbers of people have come to accept that Israeli agriculture cannot continue to receive the favoured treatment that it has enjoyed for so long. Even though some reallocation of water from agriculture to urban/industrial use has taken place, the fact remains that over 60 percent of all water is still used for irrigation in Israel. The significance of this is brought into sharp focus when it is realized that agriculture, and this includes rain-fed agriculture as well, only contributes around 2 percent to the country's GDP (Beaumont 2000b). This problem is at last being addressed as the Ministry of Agriculture now assesses the efficiency of water use in terms of the crop income per unit of water used. The hope is that high water-using crops with only low returns, such as cotton, can be phased out of production.

The high consumption of water by agriculture has also led to serious environmental problems. Perhaps, the most important is nitrate pollution of groundwater in the coastal aquifer, caused by the intensive use of fertilizers in irrigated agriculture. Salinity levels in the coastal aquifer have also increased as a result of irrigation. The problem of salinity build-up is also made worse by the fact that waters imported from Lake Kinneret to the coastal region have relatively high dissolved solid contents, as well as by the growing use of reclaimed wastewaters for irrigation purposes.

Although agriculture has been a main cause of water problems in Israel, it is important to realize that the growing urban centres have had an increasing impact in terms of the water pollution they have produced. Urban wastewaters have been responsible for polluting aquifers and for causing serious coastal pollution. The main reason for this was the lack of wastewater treatment or inadequate treatment. To a large extent, this has now been controlled by legislation, of which the Local Authorities (Sewerage) Law, 1962 and the Prevention of Sea Pollution from Land-Based Sources Law, 1988, are the most important.

The solution for water shortage in Israel, in the long-term, lies in desalination of seawater. Israel has been surprisingly slow in moving in this direction to meet its water needs, most likely for political reasons. However, in 1999, the Israeli government did agree that the Ministries of National Infrastructure, Agriculture and Finance should begin planning a major desalination facility. As desalinated water would be too expensive for use by agriculture, it shows that the government has now realized the need to supply urban regions with an uninterrupted water supply, if economic growth is to continue.

6.5.4 Jordan

Of all the countries of the Middle East, it is Jordan that faces the greatest water problems, owing to a low average precipitation and a rapidly growing population (Beaumont 2002). Aquifers are being over-pumped, especially in the south of the country, and water rationing in Amman and other large cities has been commonplace since the early 1990s. In Amman, the water distribution system has been inadequately maintained for many years and water losses through leakage can account for half the water supplied in some areas. Parts of the water administration system remain less efficient than they should be and wastewaters are often not fully treated. As a result, they can pollute potential water supplies.

The most important institution dealing with water in Jordan is the Ministry of Water and Irrigation (USAID 2001). This was established in 1988 and it brought together the already existing Water Authority of Jordan and the Jordan Valley Authority. The Minister of Water and Irrigation has overall control, with Secretaries General in the Ministry of Water and Irrigation, Water Authority of Jordan and the Jordan Valley Authority reporting directly to him. The main tasks of the Ministry of Water and Irrigation are to formulate water policy, to undertake strategic planning and resource development programmes, to formulate water allocation options, to provide a water resources data base, and to monitor and control water quality.

The Water Authority of Jordan had been established in 1984 as an independent body responsible to the Prime Minister before becoming part of the Ministry of Water and Irrigation. It is responsible for municipal and industrial water supplies and wastewater management. It plans, implements and operates all water supply and wastewater projects in Jordan.

The Jordan Valley Authority is an older organization that was created in 1977 in order to ensure the social and economic development of the lands of the Jordan Rift Valley through the management of the land and water resources of the region. From 2001, the Jordan Valley Authority is also responsible for tourist development in the area.

It is important to realize that many other parts of the government have a role to play in the management of water resources in the Kingdom. The most important is the Council of Ministers that becomes involved with water at the highest level through policy initiation, legislation and finance. The various Ministries have more specific roles. The Ministry of Planning reviews all plans put forward by the Ministry of Water and Irrigation and then liases with potential funding agencies. The Ministry of Finance oversees the budget concerning water projects and handles the financial aspects of any loans or international finance for water projects. The Ministry of Health is responsible for drinking water standards and ensuring that wastewater facilities comply with the necessary standards and regulations. It is, however, the Ministry of Agriculture that has the most important secondary role in water policy as this Ministry has the authority, under the Agriculture Law No. 20 of 1973, to exploit surface water resources through the construction of small dams to provide water for the growth of crops for animal feed. The Ministry can also drill wells to provide water for livestock. The General Corporation for Environmental Protection was established in 1995 in an attempt to bring all issues to do with the environment under one organization. Today, it is responsible for co-ordinating all environmental policies of the government in the Kingdom.

Besides the government departments, there are also a series of other organizations that play a role in the water planning process. Most of these are academic in nature and rarely have a direct executive role in water resources development. These include the Water and Environment Research and Study Centre, University of Jordan; the Environmental Research Centre, Royal Scientific Society; and the Strategic Environment and Water Resources Research Unit, Al Al-Bayt University. A non-governmental organization is the Water Conservation Association, which was established in 2000. Its role is to increase awareness of water conservation and to assist in the implementation of policies to bring such conservation about.

The basic problem for these institutions is to devise a policy so that economic growth in the country can continue at a time of increasing water shortage (Salameh and Bannayan 1993; Beaumont 2002). Currently, all available water resources are being utilized. It is only the continued mining of groundwater in the south that allows the country to survive. However, this is not a long-term solution and Jordan will have to face a series of difficult choices. Its water resources are so limited that even if it were to utilize all irrigation water for urban/industrial use, this would be insufficient to meet potential needs by the end of the next two decades. If it is to supply its cities with adequate water resources, Jordan must embark on a water desalination programme using water from the Gulf of Aqaba. The chief difficulty with this is that the water must be pumped up more than a 1,000 m to the main population centres in the northern highlands near Amman. Such pumping would be highly expensive, but it would mean that Jordan was in control of its future water needs. Other possibilities might be to import water from the Lebanon or to enter into a joint desalination programme with Israel. With both of these proposals, though, Jordan would not be able to guarantee the supply of this water in times of political instability.

6.5.5 Turkey

The modern state of Turkey dates from the 1920s with the coming to power of Kemal Ataturk. It is a secular state, although the majority of its citizens follow the Muslim faith. From its earliest years it has possessed a highly centralized government apparatus. A key feature of this bureaucratic structure has been the five-year development plan. The current five-year plan is the Eighth Plan which covers the period from 2001 to 2005.

Under the Turkish constitution, water resources are regarded as the natural wealth of the country. These are controlled by the state for the public benefit. In the Civil Code, water resources are considered under the two categories of common waters or private waters. However, private waters are relatively unimportant and most water resource development is undertaken by the state. Unlike in many other countries, groundwater use rights cannot be transferred or sold in Turkey (DSI 2002).

In Turkey, three levels of water resource management can be identified. At the highest level is broad policy and decision-making. This is carried out at Prime Ministerial level, in cooperation with the Ministry of Development, which is in effect the State Planning Organization. The next level is the executive level where detailed policies are drawn up within the confines dictated by the highest state-wide plans. It is at this level where most of the crucial decisions are taken with regard to specific projects. The final level can be thought of as the user's level, where the water is actually utilized to achieve a specific objective such as the irrigation of a crop.

At the executive level, four main organizations are responsible for the development of water resources in Turkey at the present time. These are the General Directorate of State Hydraulic Works (DSI); the General Directorate of Rural Services (GDRS); the General Directorate of the Bank of the Provinces (Iller Bank), and the General Directorate of Electric Power Resources Survey and Development Administration (EIE).

The most important organization for water resources development and management is the General Directorate of State Hydraulic Works (DSI). This was established in 1953 and now forms part of the Ministry of Energy and Natural Resources. The central offices of DSI are located in Ankara, but there are also 26 Regional Directorates throughout Turkey. DSI's responsibilities include data collection, flood control, irrigation, domestic and industrial water supply for urban regions of more than 100,000 people, hydroelectric power generation and all matters dealing with groundwater. In effect, DSI plans and manages all major aspects of water resources development.

The General Directorate of Rural Services (GDRS) is responsible for rural infrastructures, including, from the water management perspective, the construction of small reservoirs, minor irrigation schemes and drinking water supply to rural settlements. It works in close cooperation with DSI. Among the duties of the General Directorate of the Bank of the Provinces (Iller Bank) are the construction of water supply and sewage treatment facilities in smaller urban settlements, together with the financing of these schemes. Once again, there is close contact with DSI.

The General Directorate of Electric Power Resources Survey and Development Administration (EIE), which was founded in 1935, is responsible for all electric power in the country. It liaises closely with DSI when hydroelectric schemes are being considered. It is, however, DSI that constructs such facilities. The EIE, like DSI, is part of the Ministry of Energy and Natural Resources.

The Ministry of Environment is responsible for the development of Turkey's environmental policy, including those aspects dealing with water resources. In future it does seem likely that the views of this Ministry might well clash with those of DSI, when new large projects are being considered.

There is also a major regional organization that plays a vital role in water resource management in Turkey. This is the South-eastern Anatolia Project Regional Development Administration (GAP). This project deals with the social and economic development of what was a poor region, with a major emphasis placed on water management for irrigation, hydroelectric power production and urban water supply. The region of the project covers the basins of the rivers Tigris and Euphrates in their courses in Turkey. The scheme will involve the construction of 22 dams, the irrigation of 1.7 million hectares of land and the annual generation of 27 billion kWh of electricity (DSI 2002). Even by world standards, this is a huge project that will have important environmental impacts.

Although the highly centralized and bureaucratic state organizations, such as DSI, have served the country well during its major development phase since the Second World War, it has been realized during the 1990s that these state-run organizations are not as cheap to operate or as efficient as privately-run schemes. Consequently, the government is now committed to ensuring that as many water projects as possible can be handed over to those who make use of the waters. Prior to the mid-1990s, DSI managed all the major irrigation projects by itself. In recent years, however, the operation and maintenance of certain irrigation projects has been transferred to other users in an attempt to increase the overall performance of these schemes. In the future, it is hoped that other types of water resource project can also be transferred to the private sector. By so doing, it is hoped that the State will be able to recover some of the very high costs associated with water infrastructure provision at the present day.

6.6 Overall Assessment of Water Institutions in the Middle East

Despite the widespread aridity that prevails over much of the Middle East, it is interesting to note from the case studies that not all countries at the present time have a Ministry of Water as an independent entity in their governmental structures. Jordan possesses a Ministry of Water and Irrigation, while Saudi Arabia has a Ministry of Agriculture and Water. Egypt too has a Ministry of Water Resources and Irrigation. In other countries, the major water institutions or responsibilities for water provision are located within other ministries. For example, in

Turkey, DSI is located within the Ministry of Energy and Natural Resources, whereas in Israel, prime responsibility for water, until recently, has been located in the Ministry of Agriculture. In Iran, the Water Resources Management Organization is part of the Ministry of Energy. This would seem to indicate that these countries still perceive that other parts of the economy are more important than water. Whether this situation will change in the future is unknown. It must be admitted, however, that successful water resource management can be achieved if water administration is located within another Ministry, provided that sufficient power is delegated to permit water management decisions to be made.

It is equally interesting to note that in many of the countries of the Middle East, water rules and regulations which were in use until the early years of the twentieth century at the regional and local level, probably had their origins many hundred years ago, often in pre-Islamic times. In many cases, these rules and regulations which had evolved over time were incorporated into, first, customary law, and then, Islamic law, with few changes. The rationale probably was that if these rules and regulations worked satisfactorily, it was not necessary to change them. Lambton (1981) also makes the point that early Islamic societies were essentially urban-based and did not possess great expertise in the management of large irrigation systems. In such cases, it seems likely that the rulers would be happy to adopt an already working system which was able to produce adequate tax revenues.

Another interesting point is that, with rapidly growing populations, a number of the countries of the Middle East are beginning to accept the fact that in order to supply sufficient water to their urban systems, water will have to be reallocated from irrigation usage. Although a number of countries have recognized this fact, very few have come up with a coordinated policy to implement it without considerable social disruption.

A key feature of many Middle Eastern cities over the last 40 years has been the development of areas of poor quality housing on the margins of existing cities. These areas are occupied by poor people, who are often recent migrants from rural areas. Many of the people in such settlements live outside the law, in a technical sense, and so do not receive the piped water supplies and sewage disposal facilities enjoyed by other urban residents. These people have to obtain water by a variety of means. Entrepreneurs are quick to supply small quantities of water by tanker, but the costs are often many times those paid by a consumer receiving piped water. In other cases, informal arrangements are made to purchase water from people who receive piped supplies, but once again, the cost is usually high. Inevitably this means that the poorest members in society have to commit a much higher proportion of their income to water supply, than their more affluent compatriots.

Water provision for these shantytowns is something the governments of the Middle Eastern countries need to tackle if they are to ensure a healthy and productive population. There is certainly a case for subsidising the installation of piped water systems in these areas by charging other urban residents higher rates for their water provision.

The last 20 years or so have seen many of the countries of the Middle East beginning to experience the serious problems of urban water pollution that were so common in Western Europe and North America during the growth of industrialization in the nineteenth century. This has often led to the establishment of Ministries of the Environment, but, as yet, few comprehensive legislative frameworks to control pollution exist. It also seems highly likely that future water projects planned by the major water institutions are likely to cause disputes with the newly-formed Ministries of the Environment. How governments will deal with such problems is unknown. It is certainly the case that at least one organization within a State should possess an overview of all environmental issues, as in the future environmental quality issues of all kinds are likely to become much more important.

6.7 References

Abderrahman WA (2001) Water demand management in Saudi Arabia. In: Faruqui NI, Biswas AK, Bino MJ (eds) Water management in Islam. United Nations University Press, Tokyo, pp 68-78

Abderrahman WA (2000) Application of Islamic legal principles for advanced water management. Water International 25: 513-518

Amery HA (2001) Islamic water management. Water International 26: 481-489

Beaumont P (2002) Water policies for the Middle East in the 21st century: the new economic realities. International Journal of Water Resources Development 18: 315-334

Beaumont P (2000a) The quest for water efficiency - restructuring of water use in the Middle East. Water, Air, and Soil Pollution 123: 551-564

Beaumont P (2000b) Water for peace in the Middle East: The sacrifice of irrigated agriculture in Israel? Water International 3: 97-112

Beaumont P (1993) Drylands - environmental management and development. Routledge, London, 536 pages

Beaumont P (1977) Water and development in Saudi Arabia. Geographical Journal 143: 42-60

Beaumont P (1971) Qanat systems in Iran. Bulletin of the International Association of Scientific Hydrology 16: 39-50

Beaumont P (1968) Qanats on the Varamin Plain, Iran. Transactions of the Institute of British Geographers 1968, N° 45, pp 169-179

Beaumont P, Bonine M, McLachlan KS (eds) (1989) Qanat, Kariz and Khattara - traditional water systems in the Middle East and North Africa. Menas Press Ltd, Wisbech, 1989, 305 pages

Bonine ME (1977) Traditional irrigation systems and practices in central Iran. Paper presented at the Festival of Popular Traditions, Isfahan, Iran, October 1977, pp 1-29

Bureau of Water Planning and Development and Resources Corporation (1978) Iranian National Water Plan – Draft Final Report – Volume X – Summary. Bureau of Water Planning and Development and Resources Corporation, Tehran, Iran, p 37

Council for Leading Islamic Scholars (CLIS) (1978) Judgement regarding purifying wastewater. Judgement No. 64 on 25 Shawwal, 1398 AH, Thirteenth Meeting of the Council of leading Islamic Scholars (CLIS) during the second half of the Arabic month of Shawwal, 1398 AH (1998). Journal of Islamic Research 17: 40-41

DSI (General Directorate of State Hydraulic Works) (2002) Water potential and development in Turkey, in http://www.dsi.gov.tr/emwis/water_context.htm

Echo of Iran, (1971) Iran Almanac. Echo of Iran, Tehran, 808 pages

Faruqui NI, Biswas AK, Bino MJ (eds) (2001) Water management in Islam. United Nations University Press, Tokyo

Izzi Dien M (1990) Environmental Islamic law, Ethics and Society. In: Engel JR, Engel JG (eds) Ethics of Environment and Development: Global Challenge, International Response. Bellhaven Press, London

Kadouri MT, Djebbar Y, Nehdi M (2001) Water rights and water trade: an Islamic perspective. In: Faruqui NI, Biswas AK, Bino MJ (eds) Water Management in Islam, United Nations University Press, Tokyo, pp 85-93

Kay PA (2000) Measuring sustainability in Israel's water system. Water International, 25: 617-623

Kenyon K (1969) The origins of the Neolithic. Advancement of Science, London, 26: 144-160

Lambton AKS (1981) Agriculture in Medieval Persia. In: Udovitch A (ed) The Islamic Middle East, Princeton, NJ, Princeton university Press, pp 283-288.

Lambton AKS (1953) Landlord and peasant in Persia. Oxford University Press, London

McLachlan KS (1988) The Neglected Garden – The Politics and Ecology of Agriculture in Iran. IB Tauris and Co Ltd, London

Ministry of Foreign Affairs, Israel (2002) The Water Law of 1959, in http://www.mfa.gov.il/mfa/go.asp?MFAHax30

Ministry of Water Resources and Irrigation, Egypt, (2002), in http://www.mwri.gov.eg/history.html

Sadr K (2001) Water markets and pricing in Iran. In: Faruqui NI, Biswas AK, Bino MJ (eds) Water management in Islam. United Nations University Press, Tokyo, pp 103-114

Safinejad J (2001) Financing the traditional farm irrigation by qanats. Water and Development 3: 98-110

Salameh E, Bannayan H (1993) Water resources of Jordan – present status and future potentials. Friedrich Ebert Stiftung, Amman, 183 pages

Soffer A (2002) Mapping special interest groups in Israel's water policy, in http://www.biu.ac.il/SOC/besa/waterarticle7.html

USAID (2001) Improved water resources management – water resources in Jordan, in http://www.usembassy-amman.org.jo/USAID/water.html

Institutions in South African International River Basins

Anthony R. Turton and Anton Earle

7.1 Introduction

The story of Apartheid is well documented. What remains relatively unexplored however, is the connection between Apartheid and water resources management. This chapter summarizes work done by the authors over the last decade, and shows how the South African State sought to mobilize water resources as an element of its economic growth potential for more than half a century. It introduces the reader to the notion of a Hydropolitical Complex in Southern Africa, clustered around specific States and specific transboundary river basins, and it shows that institutional development in those basins is very sophisticated. In a supreme twist of irony, the evolution of institutions in South African transboundary rivers is a legacy of the Apartheid-era policy that sought to develop what became known as a Total National Strategy in response to the prevailing threat perception of the time, which was defined officially as being a Total Onslaught. This chapter concludes with an assessment of institutional development by mapping out the fundamental dynamics of a win-win and a win-lose outcome as they pertain to the South African experience.

7.2 The Southern African Hydropolitical Complex

The patterns of political interaction that occurred within the Southern African region were closely associated with Cold War rivalries. An analysis of the specific patterns of amity and enmity between States, factions, interest groups and other important actors over the last half-century reflects a predictable pattern of Cold War rivalry between America and the former Soviet Union, enacted through their respective surrogate forces. This is an element of what has been called 'overlay', which is defined as that condition when great power interests transcend mere political penetration, and come to dominate a region so heavily that the local pattern of security relations virtually ceases to operate, such as occurred with the European colonization of Africa (Buzan 1991:188; Buzan & Wæver, forthcoming). It can therefore be argued that three distinct levels of overlay existed in Southern Africa. Colonialism was the first and most written about; the Cold War was the second and probably the least written about; while Apartheid was the third, with considerable literature on this subject in general, but nothing on this subject as it pertains to water resources management. These three levels of overlay consequently had a marked impact on the politics of Southern Africa.

Building on to this notion, Buzan (1991:210) has noted the existence of a set of what he calls "regional security complexes" in the developing world, with one of these being the Southern African Regional Security Complex consisting of the Republic of South Africa, Namibia, Botswana, Zimbabwe, Angola, Zambia, Malawi, Mozambique, Lesotho and Swaziland. Buzan defines a security complex as being "a set of units (typically States) whose major processes securitization, desecuritization, or both are so interlinked that their security problems cannot reasonably be analysed or resolved apart from one another" (Buzan et al. 1998:201). In this regard, securitization is constituted by the inter-subjective establishment of an existential threat within any sector (military, political, economic, societal and environmental) with a saliency sufficient to have substantial political effects (Buzan et al. 1998:25); whereas desecuritization refers to the shifting of specific, strategically important issues out of the emergency mode and into the formal bargaining processes of the political sphere (Buzan et al. 1998:4). Security complexes thus emphasize the interdependence of both rivalry and shared interests (Buzan 1991:190), or Stated differently, reflect the shifting patterns of amity and enmity over time (Buzan 1991:198). Security complexes are analytical entities consisting of units displaying distinct patterns of both amity and enmity, characterized by predominantly inward looking national security relationships, surrounded by a zone of relative indifference.

This definition evolved from the earlier work by Buzan (1991:190) that was cited by Schultz (1995:97) in the context of what he calls the "Tigris and Euphrates Hydropolitical Security Complex". Schultz' work is the first to have linked Buzan's views on security complex theory to the study of hydropolitics. In his seminal work on the hydropolitics of the Middle East and North Africa (MENA) region, Allan (2000:245) notes the value of security complex theory, but concludes that while it explains what has happened, it does not predict what will happen in the future (Allan 2000:248). There is consequently the need for the improvement of theory if it is to have both explanatory and predicative capabilities.

Encouraged by this, the authors decided to develop a deeper understanding of how security complex theory could be applied to Southern African hydropolitics. Buzan (1991:187) notes that security is relational, meaning that one cannot understand the security of one State without understanding the international pattern of security interdependence in which it is embedded. This means that two things are relevant to a study of hydropolitics at the international level in the context of Southern Africa. Firstly, overlay is a key factor as it has shaped those patterns of amity and enmity over time. Secondly, States are locked together by virtue of their co-riparian status in any given international river basin. This has prompted the author to explore the linkage between these two aspects further (Turton 2003a, 2003b, 2003c), which has resulted in the development of a set of concepts that serve to explain some of the institutional developments in transboundary river basins that are shared between South Africa and her respective neighbours. The authors have concluded that a "Hydropolitical Complex" is emerging as a distinct component of what Buzan (1991:210) has described as the "Southern African Security Complex" (Turton 2003a, 2003b, 2003c). This Southern African Hydropolitical Complex consists of the political interaction between States that are clustered on various international river basins and comprises four key concepts:

Pivotal States: There are four pivotal States within the Southern African Hydropolitical Complex - the Republic of South Africa, Botswana, Namibia, and Zimbabwe. The hydropolitical criteria for the pivotal State label relate to two critical variables, and in order to qualify as being pivotal, a country needs to meet both of these criteria:

- A high level of economic development within each country[1];
- A high reliance on shared river basins for strategic sources of water supply as a critical element of its national economic development.

Pivotal Basins: There are two pivotal river basins within the Southern African Hydropolitical Complex - the Orange and the Limpopo. The hydropolitical criteria for the pivotal basin label relate to two critical variables, with both criteria needing to be met in order to qualify. These are:

- The strategic importance of the river basin to any one (or all) of the four pivotal States in terms of the economic development that it supports;
- The extent to which the river basin is closed, with "closure" referring to a river basin with no utilizable outflow of water (Seckler 1996) or with no more water left to be allocated to productive activities (Svendsen et al. 2001:184).

Impacted States: There are at least seven impacted States in the Southern African Hydropolitical Complex - Angola, Mozambique, Swaziland, Lesotho, Zambia, Malawi, and Tanzania. The hydropolitical criteria for the "impacted State" label relate to two critical variables, with both criteria needing to be met in order to qualify. These are:

- The level of need for water from either a pivotal basin or an impacted basin as a foundation for the economic development of the impacted State;

[1] Military power is an important indicator in this regard, because economic development can translate into the capacity to project power outside of the borders of the country. All four pivotal States have this capacity, and have demonstrated their willingness to use it. The Republic of South Africa and Botswana became involved militarily during the SADC sponsored Operation Boleas in Lesotho, which started on 22 September 1998 and ended over a month later. Both Zimbabwe and Namibia have troops deployed in the Democratic Republic of Congo (DRC) at the time of writing.

- The degree to which current (or future) economic development in an impacted State has, or is likely to be, limited by a pivotal State by virtue of the co-riparian status of the two (or more) States within a given river basin.

Impacted Basins: There are at least seven impacted basins in the Southern African Hydropoltical Complex - Zambezi, Kunene, Okavango, Incomati, Maputo, Pungué, and Save. The hydropolitical criteria for the "impacted basin" label relate to two critical variables, with both criteria needing to be met in order to qualify. These are:

- The extent to which a pivotal State relies on the water from the impacted basin for current (or future) economic development;
- The degree to which the development options of the impacted State within the impacted basin have been, or are likely to be, limited by the actions or plans of the pivotal State.

Both Impacted States and Impacted Basins can be considered as dependent variables, presenting as symptoms in the cause-effect linkage of the overall hydropolitical equation; while Pivotal States and Pivotal Basins can be considered as independent variables, acting as fundamental drivers or foci of hydropolitical dynamics. This is a useful way of understanding the patterns of amity and enmity between various States in Southern Africa, and as such it becomes a way of providing a more nuanced analysis of international relations as they pertain to the water sector. Stated differently, if international relations are how States behave under conditions of structural anarchy, and if transboundary river basins link States in a physical relationship, what can we learn from studying institutional development in those basins from a hydropolitical perspective?

7.3 History of Development in the South African Water Sector

The Southern African region as a whole is characterized by significant maldistribution of water resources, with large swathes of land receiving less than 500 mm of precipitation per annum. In fact, around 60 percent of the total mean annual runoff (MAR) of the entire Southern African region arises from just 20 percent of the land surface area. Coupled with this is an extremely high evaporative demand, which means in effect that what water does fall as rain, is almost immediately lost to evaporation. In South Africa for example, the annual average rainfall is 487 mm, with one of the lowest conversions of rainfall to runoff in the world. In fact, the total average runoff (that portion of rainfall that is not lost to evaporation and which eventually finds its way into rivers) is only some 10 percent of total annual rainfall (Rabie and Day 1992:647). Of the resultant runoff that becomes stream flow, a mere 60 percent (Rabie and Day 1992:647) to 62 percent (O'Keefe et al.

1992:278) can be economically exploited, because of the extreme variability of these rainfall events. This natural climatic variability has become a stimulus for the construction of dams in an attempt to retain as much stream flow as possible. The World Commission on Dams report listing the top ten countries by virtue of the number of dams constructed for particular purposes (irrigation, water supply, flood control and hydropower) contains South Africa and Zimbabwe (WCD 2000:373), both of which are Pivotal States.

South Africa is the most industrialized State in Southern Africa. It is also one of the most water stressed countries, with water availability acting as a possible limiting factor to its long-term economic growth potential. This makes water resources management a strategic issue. For this reason, early reconnaissance work was done with a view to assessing the future viability of sourcing water outside of the areas of need, and if necessary, in neighbouring States. There are many examples of this. A start was made in 1956 to study the hydrology of Basutoland (pre-independence Lesotho) as a possible future strategic supply of water for the South African Goldfields and related industrial complex (Ninham Shand 1956). This subsequently became the foundation of the Lesotho Highlands Water Project (LHWP) (Carter 1965; Young 1961), which is one of the largest water transfer schemes in Africa.

The water, economic development and energy nexus can be traced back to early agreements between South Africa (in its capacity as the most economically developed State in the region, but also partially in its capacity as the UN Mandated trustee of what was to become Namibia), and Portugal (in its capacity as the colonial power in both Mozambique and Angola). In this regard, an agreement was reached in 1926 for the study of the potential for hydroelectric power in the Cunene River (Treaty 1926), which is the border between what was then the Portuguese colony of Angola and the UN Mandated Territory of South West Africa under South African control. The construction of the Kariba Dam[2] in the 1950's as a cornerstone of the economic development of Rhodesia, which at that time was seen to be a bastion of European (meaning White) rule in what was then the Federation of Rhodesia and Nyasaland[3], provided renewed impetus to this water, economic development and energy nexus. The thinking at the time was that if Kariba Dam could be so successful in launching an aggressive policy of economic development in Rhodesia, then similar developments could be used to launch the economic development of South Africa and Mozambique. Consequently, an agreement was reached in 1964 to explore these issues in greater detail. This resulted in

[2] Kariba Dam was seen to be the strategic powerhouse of the Rhodesian economy. In keeping with the blatantly racist policies of the time, the hydroelectric turbines were placed on the Rhodesian side of the Zambezi River, effectively making Northern Rhodesia (what later became Zambia) totally dependent on the goodwill of the white minority government in what was then known as Southern Rhodesia (subsequently Zimbabwe).

[3] This became Malawi at independence on 6 July 1964.

a specific agreement being finalized in 1969 on the development of the Cabora Bassa Project in Mozambique downstream of Kariba Dam (Treaty 1969a), with a similar specific agreement being reached simultaneously on the development of the Cunene River (Treaty 1969b).

Other grandiose water transfer schemes were also developed. An early study on the transfer of water from the Zambezi/Chobe confluence via Botswana to South Africa (Borchert and Kempe 1985; Conley 1995; Williams 1986) was refined to a plan involving the construction of a 1,116 km concrete structure feeding $2{,}500 \times 10^6 m^3$ of water into the Vaal River Supply Area (Borchert 1987; Scudder et al. 1993:268). A feasibility study of this proposal found that the cost of water delivered to Pretoria was competitive with existing water supply schemes (Scudder et al. 1993:263; Midgley 1987). Contemporary literature refers to this as the mooted Zambezi/Limpopo link (Heyns 2002:167). It can therefore be seen that considerable energy was spent on developing plans for the sourcing of water outside of the areas of need as a future strategic supply on which the economic growth and development of the South African economy could be based.

Returning back to the notion of overlay that was introduced earlier on, three things happened that impacted on these grandiose schemes. The first was the demise of colonialism, the second was the cessation of Cold War hostilities, and with that the collapse of the Apartheid State. Let us examine the build up of these in some detail, in order to assess the impact of their subsequent demise.

Colonialism had a major impact on the politics of Africa in general, and South Africa in particular. British Prime Minister Harold Macmillan made his now famous 'Winds of Change' speech in the Cape Town Parliament in which he spelled out the strong desire for independence that was emerging in the colonies (Geldenhuys 1984:11). This desire for independence was being felt acutely in Southern Africa. The Portuguese colonies of Angola and Mozambique were both theatres for a war of liberation. The British colonies of Southern Rhodesia[4], Northern Rhodesia[5], Nyasaland and South Africa also saw various forms of anti-colonial struggle taking place. Significantly, South Africa had strong independence sentiments based on the deep-seated anger that was still being harboured by the Afrikaners at the humiliation of the Boer War defeat half a century earlier, and still in the living memory of many a survivor of the British concentration camps. The three British

[4] This became Zimbabwe on 18 April 1980, after a protracted struggle against the white minority government of Ian Smith, who declared independence from Britain unilaterally in the 1960's and subsequently repressed any form of black political expression in what became known as the "bush war".

[5] This became Zambia on 24 October 1964, and promptly became home for the liberation movements fighting white minority rule in Rhodesia and South Africa.

Protectorates in the region, Bechuanaland[6], Basutoland[7] and Swaziland were never colonies, but were functioning monarchies that had appealed for the protection of the British Monarch during the time of Zulu national expansion and Boer settler land invasions a century earlier.

The domino effect of what subsequently occurred can be traced back to two distinct trigger events. The first trigger event was undoubtedly the British withdrawal from the various colonies in the 1960's. This was partly because Britain had been financially ruined by World War II and was no longer in any state to maintain the colonies. This served to reduce the impact of colonial overlay, striking deep-seated fear into the hearts of various white minority governments who saw themselves as being bastions of European styled "civilization" in the far flung outreaches of Africa. The spate of newly-independent countries also gave new impetus to the various liberation movements operating at the time, and in particular gave them access to safe bases from which they could train their guerilla fighters, and generally regroup. The second trigger event was the Sharpeville Massacre inside South Africa, which occurred on 21 March 1960 and resulted in the death of 69 people and the wounding of a further 180 (Spitz and Chaskalson 2000:7). This event resulted in a declaration of a State of Emergency, the banning of the various liberation movements[8] and the subsequent launch of what became known as the "armed struggle". Central to this armed struggle was the search for safe rear bases, many of which were made available in the newly independent countries of the region, from which guerilla attacks could be launched.

If these were the two trigger events then a whole series of consequences flowed from them. One of these took the form of a deep-seated fear of independence - a sentiment that was expressed mostly by the White minority, who increasingly felt insecure in the face of the so-called "Winds of Change". A dramatic manifestation of this fear was the Unilateral Declaration of Independence (UDI) for Rhodesia by Prime Minister Ian Smith, which resulted almost immediately in a repressive bush war. Afraid as they were of the political consequences of independence in the face of the events surrounding the Sharpeville Massacre, South Africa was quick to support Smith's UDI from Britain. This polarized the Southern African region, and set out the battle lines for the next quarter century of struggle. South Africa deployed paramilitary units in support of Smith. This gained momentum in 1966 when the guerilla war was launched in Namibia (Frankel 1984), making the Caprivi Strip a theatre of war in which South Africa and Rhodesia had a strategic interest. Both South Africa and Rhodesia deployed forces in support of the Portuguese army in Angola and Mozambique. The Portuguese forces were taking a battering, however, and similar to the American experience in Vietnam, the trickle

[6] This became Botswana on 30 September 1966.

[7] This became Lesotho on 4 October 1966.

[8] The imprisonment of Nelson Mandela and others occurred as a direct result of this event.

of dead and wounded back to Portugal was met by growing public resentment. In 1975 a *coup d'état* took place in Portugal, overthrowing the military regime and changing the balance of power in Southern Africa overnight. Portugal decided to withdraw from all of its former colonies, leaving no transitional governments in place, and in some cases leaving different factions to fight it out in order to determine who would become the new legitimate government of the now independent colony. The effects of this were startling, coming as they did on top of already heightened fears about Black majority rule. In Rhodesia, a second military front was opened up, with guerilla forces finding a safe haven in the now liberated Mozambique. From that moment on, a military solution was no longer possible in Rhodesia even with increased South African support. In Angola, three factions that had been united in their struggle against Portuguese colonial oppression, turned against each other in a bitter civil war that was to last until 2002. This sudden turn of events saw South African military forces being deployed in Angola, where they would stay until their eventual defeat at the Battle of Cuito Cuanavale in 1988 (Simon 1991:187).

So how is this all relevant to institutional development in the water sector?

The two trigger events noted above, and in particular the deteriorating security situation that arose from these two events, gave birth over time to a specific threat perception in South Africa that was based on the "carrot" of development cooperation as an inducement to the non-support of the various liberation movements, and the "stick" of military retaliation in the event that this positive inducement was not heeded. The first discernible elements of this can be traced back to the 1973 White Paper on Defence, which introduced the concept of a "total strategy" into the South African political psyche for the first time (Geldenhuys 1984:140). The 1977 Defence White Paper was largely devoted to refining the concept of a total national strategy, first mooted in 1973, as an official policy. This defined a "total national strategy" as being "the comprehensive plan to utilize all the means available to a State according to an integrated pattern in order to achieve the national aims within the framework of specific policies" (Geldenhuys 1984:140). This total strategy had its roots in the counter-revolutionary experiences of the Americans in Vietnam, the British in Malaya, and the French in Algeria and Indo-China. The term 'total strategy' is derived directly from André Beaufre's work *An Introduction to Strategy* (Frankel 1984:46) As such it resonated well with the security elites in the emerging South African garrison State, with its peculiar threat perception that interpreted the Cuban and East German support of the African liberation movements in Southern Africa, as being evidence of a 'total onslaught', driven by Soviet imperialism (Frankel 1984:55).

This saw the subsequent development of a two-pronged approach to security-related issues, and heralded the start of the gradual securitization of water resources management. The one element was based on a strong military response to any threat, supported by destabilization via economic means (Gutteridge 1983:38). The olive branch of economic development thus became securitized, with far-reaching ramifications that are being felt today. This was given greater structure when P.W. Botha[9] produced a 12-point plan for survival at the Nationalist Party[10] Congress in 1979 (Gutteridge 1985a:93). Central to this total strategy was economic development and the resultant dependencies that would emerge from this. The foundation for this thinking can be traced to the speech made by Prime Minister Vorster in 1974, in which he spoke of a "power block" of States (Geldenhuys 1984:39). This was subsequently refined when he spoke of a constellation of politically independent States maintaining close economic ties. When P.W. Botha came to power, he used what he called a Constellation of Southern African States (CONSAS) as the basis of his policy (Geldenhuys 1984:41). Foreign Minister Pik Botha subsequently announced in 1979 that this vision embraced some 40 million people south of the Cunene and Zambezi Rivers all joining forces to design a common approach to ensure economic and political security. The grandiose water transfer schemes from the Zambezi noted above (Scudder et al. 1993:263; Midgley 1987:15), should consequently be evaluated in this light, because the water, economic development and state security nexus was a fundamental driver of that process - a fact that is totally ignored in all of the contemporary literature on South African institutional development in the water sector.

At the Lancaster House Conference in late 1979, the foundation was laid for the cessation of hostilities in Rhodesia. Bishop Abel Muzorewa was widely tipped to win the elections, so it came as a great shock to the security elites in Pretoria when Robert Mugabe swept to victory in 1980. Mugabe immediately announced that he had no intention of joining the proposed CONSAS. Instead Zimbabwe, along with Botswana, Lesotho, Swaziland, Mozambique, Angola, Zambia, Malawi and Tanzania, joined forces in the Southern African Development Coordination Conference (SADCC), which was formally launched in Lusaka during 1980. This new grouping was specifically designed to reduce their combined dependence on South Africa (Baynham 1989:88; Conley & van Niekerk 1998:145), and was quickly dubbed the 'counter-constellation' (Geldenhuys 1984:41). The establishment of SADCC was thus a direct response to South Africa's policy of destabilization. The linkage between water and development became manifest at the 4th SADCC consultative conference which was held in Lusaka during 1984. Opening the conference, President Kaunda of Zambia said that the effects of water scarcity and drought had resulted in food deficits and poor prospects for agricultural development in Southern Africa (Africa 1984).

---

[9] P.W. Botha became the first Executive President of South Africa in 1984 after the constitution was changed to allow for the concentration of all political power in this one position.

[10] The Nationalist Party was the ruling political party from 1948 until 1994 when the first democratic elections saw Nelson Mandela become the first Black President of South Africa.

The emergence of this total strategy approach saw South African foreign policy becoming captive to the State Security Council (SSC) (Frankel 1984), which had an all-consuming security focus to it. Seen in this light, every aspect of foreign relations became securitized, including cooperation over water resources. An example of the impact of the so-called total strategy in the water sector can be found in a paper written by the Chief Engineer of the Rand Water Board, who used the concept to conceptualize the need for the South African economic heartland to gain access to secure supplies of water (James 1980). Listed in this document are inter-basin transfer (IBT) schemes such as the Lesotho Highlands Water Project (LHWP), the Tugela-Vaal link and the mooted Okavango development. Significantly, gaining access to the Okavango is referred to in the context of CONSAS (Blanchon 2001:123), thereby establishing water as a strategic element of this overall policy.

The first active military retaliation was launched in 1981 (Geldenhuys, 1984:140), with a South African Defence Force (SADF) Special Forces raid on guerilla bases at Matola near Maputo in Mozambique (Gutteridge 1981:14). This was followed in 1982 with retaliatory attacks against guerilla targets in Maseru, Lesotho (Gutteridge 1983:35). These signalled that South Africa was not prepared to countenance what they perceived as being terrorist or guerilla bases in neighbouring States (Geldenhuys 1983:47) using rhetoric that resembles the contemporary American-led "War on Terror". This was manifest in subsequent attacks on targets in Angola, Mozambique, Lesotho, Botswana and Zimbabwe. A Southern African Defence Zone was conceived embracing Namibia, Botswana, Swaziland and Zimbabwe to counter the presence of East German troops in Angola and Mozambique (Gutteridge 1981:19). This marked an escalation in South African destabilization tactics, involving both military action and economic pressure (Geldenhuys 1983:43), which simply strengthened SADCC resolve to liberate their members from the stranglehold of South African economic power (Geldenhuys 1983:47). South Africa increasingly became isolated (Geldenhuys 1990) as a pariah State, associating closely with the experiences of Taiwan and Israel (Frankel 1984:65). Central to this was the deeply held belief by government decision-making elites that these three States were strategic pillars against a global Marxist onslaught that the Free World would not be able to ignore in the long-term (Geldenhuys 1984:116), so the actions were designed to buy time until that realization was made.

On the "carrot" side of this strategic equation, significant peace inducements were offered. A security agreement between South Africa and Swaziland was reached (Treaty 1982), which was immediately supported by an economic cooperation agreement (Treaty 1983a). This was followed shortly afterwards when the Nkomati Peace Accords (Treaty 1984a) were signed by President Samora Machel of Mozambique and Prime Minister P.W. Botha of South Africa, during March 1984 (Gutteridge 1985a:94). Water was intimately linked to the Nkomati Accord process when an agreement was signed during May in Cape Town between Mozambique, Portugal and South Africa on the revival of the Cabora Bassa Project

(Treaty 1984b). Similar security agreements were mooted between South Africa and Botswana, where economic cooperation and possible access to the Okavango River was discussed; and Lesotho, where access to water was also a feature (Gutteridge 1985a:100).

The need for such security agreements was underscored by unrest within South Africa that was escalating uncontrollably, with the SADF being committed to internal riot control. The African National Congress (ANC) held a high-level meeting in Kabwe, Zambia during June 1985 at which time a decision was taken to allow attacks on soft targets (Gutteridge 1985b:129). Special Forces reprisal was launched in Kabwe a few days later. A State of Emergency was announced in 1985, giving security forces wider powers (Gutteridge 1985b:124). The ANC leadership started to regard the internal situation as a 'peoples' war' from this moment onwards (Gutteridge 1985b:130). The situation deteriorated rapidly with a flight of foreign capital threatening a total collapse of the economy, so all foreign currency trading was suspended in South Africa on 27 August 1985 (Gutteridge 1985b:144). The security problems were grave indeed, with South Africa literally teetering on the very brink of financial collapse.

During 1986 violence erupted in Natal between *comrades* from the ANC and Zulu *impis* from Chief Buthelezi's Inkatha Freedom Party (IFP) (Gutteridge 1990:168). This degenerated into a localized low-intensity civil war, which endured until after the election of Nelson Mandela as the first democratic President of South Africa (Percival and Homer-Dixon 1995:3). In neighbouring Lesotho, Major General Justin Lekhanya overthrew Leabua Jonathan during a military *coup d'état* on 20 January 1986 (Lawrence 1986). Shortly after this the Treaty on the Lesotho Highlands Water Project was signed on 24 October 1986 between 'Pik' Botha of South Africa and Col. Thaabe Letsie of Lesotho (Treaty 1986a), fuelling speculation about possible South African involvement (Homer-Dixon 1994:19). Commentary on this project from that time reflects the socio-economic benefit angle that was central to the total strategy approach (Vorster 1988:95). It is against this background that the feasibility of transferring water from the Zambezi via Botswana to Pretoria (Scudder et al. 1993:263; Midgley 1987:15) should be evaluated. The existing transboundary water supply from the Molatedi Dam in South Africa to Gaborone should also be seen in light of this total strategy approach.

7.4 An Assessment of Post-Apartheid Institutional Development

Having noted the fundamental driving forces at work in Southern Africa, it now remains to determine how these have impacted on institutional development in the South African water sector. A detailed assessment of the four international river basins to which South Africa is a riparian State has shown that two distinct categories of outcome have emerged as a result of the political dynamics noted above (Turton 2003a). For ease of assessment these can be called plus-sum (win-win) outcomes and zero-sum (win-lose) outcomes.

If international relations refer to the way that sovereign States interact in an overall system of structural anarchy, then it is encouraging to note that anarchy is what States choose to make of it. Regimes are a recognized way of ordering that anarchy, by reducing uncertainty and institutionalising relationships between States. In short, a regime is a key element in the desecuritization of issues such as water resources management in transboundary rivers. A regime is defined as a set of implicit or explicit principles, norms, rules and decision-making procedures around which actors' expectations converge in a given area of international relations (Krasner 1982:186, 1983:2). Regimes differ from institutions because they lack enforceability, but they are a specific form of institution (Turton 2003a).

7.4.1 Plus-Sum Outcomes

Plus-sum outcomes occurred in the Orange and Limpopo River (Pivotal Basins), and in the Incomati and Maputo Rivers (Impacted Basins). In this regard, South Africa as the regional hegemonic power that was driven by a deep-seated sense of insecurity, sought to limit the range of actions left open to various neighbouring States in terms of the Total National Strategy approach of the Apartheid years. The one element of this two-pronged approach was to offer the inducement of economic development as an incentive to the respective co-riparian States not to harbor guerilla forces on their territory. This "carrot" of development took the form of major hydraulic installation along with the necessary support infrastructure such as roads, bridges, hospitals, schools and communication networks. Available evidence suggests that when two specific sets of conditions were met by the non-hegemonic State when they were offered the incentive of economic cooperation by South Africa, then the outcome became plus-sum in configuration. These two conditions were:

- When the co-riparian State viewed the incentive that was offered by South Africa purely in terms of national self-interest. Stated simplistically, this means that the negotiators asked themselves what their country could gain from the deal being offered.
- When the co-riparian States chose to frame the problem being confronted in non-ideological terms.

This set of conditions was met by Lesotho in the Orange River, which gave rise to a series of agreements. In 1978 a Joint Technical Committee (JTC) was established between South Africa and Lesotho to investigate the feasibility of the Lesotho Highlands Water Project (LHWP). This led to the formal signing of an agreement in 1986 to establish the LHWP, along with various parastatal bodies needed to complete the project, under the overall management of the Joint Permanent

Technical Commission (JPTC) (Treaty 1986a). This was upgraded later to accommodate the closer cooperation between the two countries in the immediate post-Apartheid period (Treaty 1999).

Namibia was a special case because it was administered as a de facto province of South Africa during the Apartheid era. Formal regime development between the two countries started with the establishment of a Joint Technical Committee (JTC) just prior to Namibian independence (Treaty 1987). This was upgraded immediately after Namibian independence but before the first democratic elections in South Africa when the Permanent Water Commission (PWC) was established (Treaty 1992a). Simultaneously, an agreement was signed establishing the Joint Irrigation Authority (JIA) to implement the Treaty on the Vioolsdrift and Noordoewer Joint Irrigation Scheme (VNJIS) (Treaty 1992b).

These two separate sets of bilateral arrangements were amalgamated in the post-Apartheid era with the establishment of the Orange-Senqu River Commission (ORASECOM) (Treaty 2000). The Orange River Basin is therefore managed jointly by all riparians, including Botswana, which is a really interesting case that deserves special mention. Botswana is a riparian to the Orange River by virtue of the Molopo River, which is endoreic and contributes no stream flow to the main stem of the river. Botswana also uses none of the water from the Orange River. Yet by virtue of its de facto riparian status, Botswana has opened up a wider range of strategic choices for itself in future, including the strengthening of bargaining positions in other river basins that are more strategically important to Botswana (Okavango and Limpopo). Significantly, Botswana is using its upstream status in the Limpopo, combined with the known strategic interest that South Africa has in possibly gaining access to the Zambezi River, as a bargaining tool. In short Botswana is offering South Africa the possibility of becoming a joint partner in any future scheme to mobilize Zambezi River water and transfer it to Gaborone in Botswana, or to Bulawayo in Zimbabwe. This strengthens the case for an emerging Hydropolitical Complex in Southern Africa as it is impacting on the international relations between States in a way that was not previously relevant.

This set of conditions was also met by Botswana in the Limpopo River, which gave rise to the establishment of the Joint Permanent Technical Committee (JPTC) in 1983 (Treaty 1983b). This functioned well from its inception and was consequently upgraded to a Joint Permanent Technical Commission (JPTC) on the Limpopo River Basin as far as it constitutes the borders between the two countries (Treaty 1989). This is functioning well and has been responsible for the Joint Upper Limpopo Basin Study (JULBS) (JPTC 1991), which is investigating a range of issues including the construction of three new dams. Bilateral relations were further strengthened in the immediate post-Apartheid period with the establishment of the Joint Permanent Commission for Cooperation (JPCC) (Treaty 1997). This is a broad inter-governmental agreement aimed at fostering better relations in a variety of fields, and not just water.

Similarly, this set of conditions was met by Swaziland on the Incomati River. During the height of the political instability a basin-wide agreement was attempted between South Africa, Swaziland and Mozambique when the Tripartite Permanent Technical Committee (TPTC) was launched (Treaty 1983c). When this failed (for reasons to be discussed under the zero-sum outcome heading below) South Africa offered Swaziland a bilateral agreement in the form of a Joint Water Commission (JWC) (Treaty 1992c). This saw the establishment of the Komati Basin Water Authority (KOBWA) (Treaty 1992d), which was responsible for the development of the Maguga and Driekoppies Dam complex - a smaller version of the LHWP without the major transboundary water transfer pipelines.

7.4.2 Zero-Sum Outcomes

Zero-sum outcomes occurred in the Limpopo (Pivotal Basin), and in the Incomati and Maputo Rivers (Impacted Basins). In this regard, South Africa's desire to limit the range of actions left open to various neighbouring States in terms of the Total National Strategy approach of the Apartheid years was frustrated by the non-interest in the agreement being offered by specific co-riparian States. This zero-sum configuration occurred when two specific conditions were met by co-riparian States:

- When the co-riparian State did not view the incentive that was offered by South Africa purely in terms of national self-interest. Stated simplistically, this means that the negotiators chose to view the offer in terms of factors other than direct material benefits to be gained by their countries if agreement were to be reached.
- When the co-riparian States chose to frame the problem being confronted in ideological terms, thereby linking the narrowly defined economic benefits to a broader defined ideological struggle between capitalism and communism.

This set of conditions was met by Zimbabwe in the Limpopo River, and by Mozambique in the Limpopo, Incomati and Maputo River basins. The TPTC agreement (Treaty 1983c) was used to manage the Limpopo, Incomati and Maputo River Basins, but Zimbabwe was left out of the negotiation process because of the high level of hostility that accompanied Zimbabwean independence as noted earlier. Mozambique, on the other hand, chose to support Zimbabwe in the so-called liberation struggle, which was against racism, colonialism and capitalism. The decision to offer this support was thus motivated almost totally by ideology, which clouded the vision of the negotiators and caused them not to view any offer of development as a legitimate offer. A few years later, attempts were made to revive the process in the form of the Limpopo Basin Permanent Technical Committee (LBPTC) involving Botswana, South Africa, Zimbabwe and Mozambique (Treaty 1986b), making this the first basin-wide agreement to be reached in Southern Africa. It failed however, and the bilateral route was followed with Botswana (refer to the Plus-Sum Outcome noted above); with Mozambique in the form of the revived TPTC in 1989; and with no agreements ever reached between South Africa and Zimbabwe.

In both of these cases - Zimbabwe and Mozambique - the co-riparian State was effectively ignored after they failed to buy into the regime being offered, and once that happened bilateral arrangements were made with the other riparian States that were willing to cooperate. The implications of this were severe for both countries, as they have been marginalized in the process, and in the post-Apartheid era at least one of these States (Mozambique) has decided to re-engage itself, but this has been from a position of gross disadvantage. This took the form of the so-called Incomaputo Agreement (Treaty 2002) that was signed during the World Summit on Sustainable Development (WSSD) in Johannesburg. To date there is no functioning agreement with Zimbabwe, and ironically relations are strained between Zimbabwe and Mozambique - its former ally in the ideological struggle - over a number of shared rivers in which Zimbabwe is acting as a typical upstream riparian with Mozambique being impacted negatively like any typical downstream riparian where no regime is in place to regulate inter-State behaviour.

7.5 Concluding Remarks

International relations between sovereigns States occur in a condition of structured anarchy. One of the manifestations of this is in transboundary river basins. Regimes are useful institutions that serve to regulate inter-State behaviour, and consequently provide some structure to that anarchy. In the case of South Africa, the impetus to regime creation was provided by insecurity in the face of a liberation struggle, which gave rise to an official government policy that chose to offer the "carrot" of cooperation in all spheres (with water resources management being one of these spheres) and the "stick" of military action. This resulted in a relatively sophisticated series of water management institutions in transboundary river basins, which has a positive impact in the post-Apartheid era of reconstruction and economic revitalization. This has given rise to the notion of a Hydropolitical Complex as a distinct component of the Southern African Regional Security Complex.

7.6 References

Africa (1984) SADCC Moves to tackle drought. In Africa 153: 74-75

Allan JA (2000) The Middle East water question: hydropolitics and the global economy. IB Tauris, London

Baynham S (1989) SADCC security issues. Africa Insight 19: 88-95

Blanchon D (2001) Les nouveaux enjeux géopolitiques de l'eau en Afrique Australe. Hérodote Revue de Géographie et de Géopolitique N° 102: 113-137

Borchert G (1987) Zambezi aqueduct. Institute of Geography and Economic Geography, University of Hamburg, Hamburg

Borchert G, Kemp S (1985) A Zambezi aqueduct. SCOPE/UNEP Sonderband Heft N° 58: 443-457

Buzan B (1991) People, States and fear, an agenda for international security studies in the post-Cold War Era. Harvester Wheatsheaf, London

Buzan B, Wæver O. (Forthcoming). Regions and powers: the structure of international security. Oxford University Press, London

Buzan B, Wæver O, de Wilde J (1998) Security: a new framework for analysis. Lynne Rienner, London

Carter CA (1965) Basutoland as a source of water for the Vaal Basin. The Civil Engineer in South Africa, 7: 217-228

Conley A, van Niekerk P (1998) Sustainable management of international waters: the Orange River case. In: Savenije H and van der Zaag P (eds) The Management of Shared River Basins: Experiences from SADC and EU. Netherlands Ministry of Foreign Affairs, The Hague, pp 142-159

Conley AH (1995) A synoptic view of water resources in Southern Africa. Paper presented at the Conference of Southern Africa Foundation for Economic Research on Integrated Development of Regional Water Resources, Nyanga, Zimbabwe, November 1995

Frankel PH (1984) Pretoria's Praetorians: civil-military relations in South Africa. Cambridge University Press, London

Geldenhuys D (1990) Isolated States: a comparative analysis. Jonathan Ball Publishers, Johannesburg

Geldenhuys D (1984) The diplomacy of isolation: South African foreign policy making. Macmillan South Africa, Johannesburg

Geldenhuys D (1983) The destabilization controversy: an analysis of a high-risk foreign policy option for South Africa. Conflict Studies N° 148: 11-26. In: Gutteridge W (ed) 1995. South Africa: From Apartheid to National Unity, 1981-1994. Aldershot, Hants & Brookfield, VT: Dartmouth Publishing, pp 42-57

Gutteridge W (1990) South Africa: Apartheid's endgame. Conflict Studies N° 228: 1-37. In Gutteridge W (ed) (1995) South Africa: From Apartheid to National Unity, 1981-1994. pp 147-182. Aldershot, Hants & Brookfield, VT: Dartmouth Publishing.

Gutteridge W (1985a) South Africa: evolution or revolution? Conflict Studies N° 171: 3-39. In: Gutteridge W (ed) (1995) South Africa: From Apartheid to National Unity, 1981-1994, pp 85-120. Aldershot, Hants & Brookfield, VT: Dartmouth Publishing.

Gutteridge W (1985b) The South Africa crisis: time for international action. Conflict Studies N° 179: 1-23. In: Gutteridge W (ed) (1995) South Africa: From Apartheid to National Unity, 1981-1994, pp 123-144. Aldershot, Hants & Brookfield, VT: Dartmouth Publishing.

Gutteridge W (1983) South Africa's national strategy: implications for regional security. Conflict Studies N° 148: 3-9. In: Gutteridge W (ed) (1995) South Africa: From Apartheid to National Unity, 1981-1994, pp 35-41. Aldershot, Hants & Brookfield, VT: Dartmouth Publishing.

Gutteridge W (1981) South Africa: strategy for survival? Conflict Studies N° 131: 1-33. In: Gutteridge W (ed) (1995) South Africa: From Apartheid to National Unity, 1981-1994, pp 1-32. Aldershot, Hants & Brookfield, VT: Dartmouth Publishing

Heyns P (2002) Interbasin transfer of water between SADC countries: a development challenge for the future. In: Turton AR & Henwood R (eds) Hydropolitics in the Developing World: A Southern African Perspective. African Water Issues Research Unit (AWIRU), Pretoria, pp 157-176

Homer-Dixon TF (1994) Environmental scarcities and violent conflict: evidence from cases. International Security, 19: 5-40

James LH (1980) Total water strategy needed for the Vaal Triangle: meeting the challenge of the eighties. In Construction in Southern Africa, pp 103-111

JPTC (1991) Joint upper Limpopo Basin study. Stage 1. Main Report. Botswana and South Africa: Joint Permanent Technical Committee on Water Affairs

Krasner SD (1983) Structural causes and regime consequences: regimes as intervening variables. In: Krasner SD (ed) International Regimes. Cornell University Press, Ithaca, pp 1-21

Krasner SD (1982) Structural causes and regime consequences: regimes as intervening variables. International Organization, 36: 185-205

Lawrence P (1986) Pretoria has its way in Lesotho. Africa Report, March-April 1986, pp 50-51

Midgley DC (1987) Inter-state water links for the future. The South African Academy of Science and Arts Symposium: Water for Survival, August

Ninham Shand (1956) Report on the regional development of the water resources of Basutoland. Report commissioned for the Government of Basutoland by the Director of Public Works under Terms of Reference 1290/W30, 11 October 1955. Ninham Shand, Cape Town

O'Keeffe J, Uys M, Bruton MN (1992) Freshwater systems. In: Fuggle RF and Rabie MA (eds) (1992) Environmental Management in South Africa. Juta & Co, Johannesburg, pp 277-315

Percival V, Homer-Dixon T (1995) Environmental scarcity and violent conflict: the case of South Africa. American Association for the Advancement of Science, Washington

Rabie MA, Day JA (1992) Rivers. In: Fuggle RF and Rabie MA (eds) Environmental Management in South Africa. Juta & Co, Johannesburg, pp 647-668

Schulz M (1995) Turkey, Syria and Iraq: a hydropolitical security complex. In: Ohlsson L (ed) Hydropolitics: Conflicts over Water as a Development Constraint. Zed Books, London pp 91-122

Scudder T, Manley RE, Coley RW, Davis RK, Green J, Howard GW, Lawry SW, Martz PP, Rogers PP, Taylor ARD, Turner SD, White GF, Wright EP (1993) The IUCN review of the Southern Okavango integrated water development project. IUCN Communications Division, Gland

Seckler D (1996) The new era of water resources management: from "dry" to "wet" water savings. IIMI Research Report N° 1. International Irrigation Management Institute (IIMI), Colombo

Simon D (1991) Independent Namibia one year on. Conflict Studies N° 239: 1-27. In: Gutteridge W (ed) (1995) South Africa: From Apartheid to National Unity, 1981-1994. pp 185-211. Aldershot, Hants & Brookfield, VT: Dartmouth Publishing

Spitz R, Chaskalson M (2000) The politics of transition: a hidden history of South Africa's negotiated settlement. Witwatersrand University Press, Johannesburg

Svendsen M, Hammond Murray-Rust D, Harmancioglu N, Alpaslan N (2001) Governing closing basins: the case of the Gediz River in Turkey. In: Abernethy CL (ed) Intersectoral Management of River Basins. International Water Management Institute (IWMI), Colombo, pp 183-214

Treaty (2002) Tripartite interim agreement between the Republic of Mozambique and the Republic of South Africa and the Kingdom of Swaziland for cooperation on the protection and sustainable utilization of the water resources of the Incomati and Maputo watercourses. Signatory document, signed by representatives of the three Governments. Johannesburg, 29 August 2002, 48 pages

Treaty (2000) Agreement between the Governments of the Republic of Botswana, the Kingdom of Lesotho, the Republic of Namibia, and the Republic of South Africa on the establishment of the Orange-Senqu River Commission. Signatory document, signed by representatives of the four Governments. Windhoek, 3 November 2000, 11 pages

Treaty (1999) Protocol VI to the Treaty on the Lesotho Highlands water project. Supplementary arrangements regarding the system of governance for the project. Signatory document, signed by representatives of the two Governments. Pretoria, 4 June 1999, 25 pages

Treaty (1997) Agreement between the Government of the Republic of South Africa and the Government of the Republic of Botswana for the establishment of a Joint Permanent Commission for Cooperation, 4 pages

Treaty (1992a) Agreement between the Government of the Republic of South Africa and the Government of the Republic of Namibia on the establishment of a Permanent Water Commission. Signatory document, signed by representatives of the two Governments. Noordoewer, 14 September 1992, 9 pages

Treaty (1992b) Agreement on the Vioolsdrift and Noordoewer joint irrigation scheme between the Government of the Republic of South Africa and the Government of the Republic of Namibia. Signatory document, signed by representatives of the two Governments. Noordoewer, 14 September 1992, 33 pages

Treaty (1992c) Treaty on the establishment and functioning of the joint water commission between the Government of the Kingdom of Swaziland and the Government of the Republic of South Africa. Signatory document, signed by representatives of two Governments. Mbabane, 13 March 1992, 17 pages

Treaty (1992d) Treaty on the development and utilization of the water resources of the Komati River Basin between the Government of the Republic of South Africa and the Government of the Kingdom of Swaziland. Signatory document, signed by representatives of two Governments. Mbabane, 13 March 1992, 51 pages

Treaty (1989) Agreement between the Government of the Republic of Botswana and the Government of the Republic of South Africa relative to the establishment of a Joint Permanent Technical Commission. Signed by representatives of the two Governments in June 1989

Treaty (1987) Cooperation agreement between the Government of the Republic of South Africa and the transitional Government of National Unity of South-West Africa / Namibia regarding the control, development and utilization of the water from the Orange River. Signatory document, signed by representatives of two Governments. Windhoek, 13 November 1987, 5 pages

Treaty (1986a) Treaty on the Lesotho Highlands water project between the Government of the Republic of South Africa and the Government of the Kingdom of Lesotho, 85 pages

Treaty (1986b) Agreement between the Government of the Republic of Botswana, the Government of the People's Republic of Mozambique, the Government of the Republic of

South Africa and the Government of the Republic of Zimbabwe relative to the establishment of the Limpopo Basin Permanent Technical Committee, 4 pages

Treaty (1984a) Agreement of non-aggression and good neighbourliness between the Government of the Republic of South Africa and the Government of the People's Republic of Mozambique. In: South Africa Treaty Series vol. 14, 1986

Treaty (1984b) Agreement between the Governments of the Republic of South Africa, the People's Republic of Mozambique and the Republic of Portugal Relating to the Cabora [sic] Bassa Project. In: South Africa Treaty Series, vol. 15; 1986

Treaty (1983a) Agreement between the Government of the Republic of South Africa and the Government of the Kingdom of Swaziland with regard to financial and technical assistance for the construction of a railway link in the Kingdom of Swaziland.

Treaty (1983b) Agreement between the Government of the Republic of Botswana and the Government of the Republic of South Africa relative to the establishment of a Joint Permanent Technical Committee. Signed by Representatives of the Two Governments in November 1983

Treaty (1983c) Agreement between the Government of the Kingdom of Swaziland, the Government of the People's Republic of Mozambique and the Government of the Republic of South Africa relative to the establishment of a Tripartite Permanent Technical Committee. Signed at Pretoria on 17 February 1983. In: South Africa Treaty Series N° 12, 1986

Treaty (1982) Agreement between South Africa and Swaziland. South Africa Treaty Series. 15: 6-7

Treaty (1969a) Agreement between the Governments of the Republic of South Africa and Portugal relative to the Cahora Bassa Project. In: South Africa Treaty Series, vol. 7

Treaty (1969b) Agreement between the Republic of South Africa and the Government of Portugal in regard to the first phase development of the water resources of the Kunene [sic] River Basin. In: South Africa Treaty Series, vol 1

Treaty (1964) Agreement between the Government of the Republic of South Africa and the Government of Portugal in regard to rivers of mutual interest and the Cunene River Scheme. In: Republic of South Africa Treaty Series. Government Printer, Pretoria, vol 7

Treaty (1926) Agreement between South Africa and Portugal regulating the use of the Kunene [sic] River for the purposes of generating hydraulic power and of inundation and irrigation of the mandated territory of South-West Africa. In: League of Nations Treaty Series LXX: 315

Turton AR (2003a) The political aspects of institutional developments in the water sector: South Africa and its international river basins. Unpublished draft of a D.Phil. Thesis, Department of Political Sciences, University of Pretoria, Pretoria, p 353

Turton AR (2003b) Environmental security: A Southern African perspective on transboundary water resources management. In: Environmental Change and Security Project Report. The Woodrow Wilson Centre. Issue 9. Woodrow Wilson International Center for Scholars, Washington DC

Turton AR (2003c) The evolution of water management institutions in select Southern African international river basins. In: Biswas AK, Unver O, Tortajada C (eds) Water and Regional Development. Oxford University Press, London

Vorster MP (1988) The Lesotho Highlands water project. South African Yearbook of International Law, 13: 95-118

WCD (2000) Dams and development: a new framework for decision-making. Earthscan, London

Williams GJ (1986) Zambezi water for South Africa? Zambia Geographical Journal, N° 36: 57-60

Young BS (1961) Projected hydro-electric schemes in Basutoland. Journal of Geography 60: 225-230

Property Rights, Water Rights and the Changing Scene in Western Water

Charles W. Howe

8.1 Introduction

As population densities and resource exploitation escalated, the intensity and variety of externalities reached a point where increasing social monitoring and supervision became worthwhile. De jure property rights became increasingly the product not only of legally recorded deeds or registries but also of regulations, tort law and customs that circumscribed the utilization of resources. Water rights are a form of property that, under western U.S. appropriations doctrine, can be bought and sold, but major externalities are created whenever those rights are transferred to new uses. Western states are increasingly taking these impacts into account in the administration of water transfers. While some critics consider this elaboration of criteria to be an erosion of property rights leading to a loss of economic efficiency, this study concludes that this trend will and should continue.

8.2 The Definition and Evolution of Property Rights

In the days of the open frontier, property rights could be very simple: population was sparse, resources were plentiful, and externalities were seldom created. People were free to use land and other assets as they wanted. A description of the boundaries of the property was sufficient to define the property right. As population density and the proximity of human activities increased, significant externalities were created and new forms of liability (tort law) came to be recognized. Collective values as well as individual values had to be protected. The more complex set of rights and obligations gradually became embodied in formally specified property rights, detailed regulations and liability rules. This process of *redefining property rights* continues today. In reality, the practical distinction among property rights, regulations and liability rules becomes very fuzzy: they mutually determine what can be done with property.

It is helpful to think of "property rights" as those legal attributes over which the owner has control, attributes that can be passed on to subsequent owners. Schlaiger and Ostrom (1993) in their work on property rights in traditional systems identify four increasingly inclusive levels of control over property: access; management of the resource; exclusion of other users; and the ability to sell or alienate the resource. Different levels of rights are defined by different mixes of custom, judicial decisions, legislation, in the tort law system as well as in formally recorded deeds. The case of access alone is illustrated by "open access" lands such as those controlled by traditional villages or communes or a tribe. Colorado Supreme Court decision in 2003 ruled that the community of San Luis in southern Colorado can have access to the Taylor Ranch for purposes of firewood gathering and animal grazing on the Taylor Ranch, a privately-owned ranch, that evolved from a Spanish land grant. Ability to manage the resource system is illustrated by tenant farming contracts which also would generally allow exclusion of others. These attributes plus the ability to sell or "alienate" the property are typical of modern ownership, although denial of access can be moderated by traditional rights (the walking paths of Europe) and management can be constrained by regulations.

Institutions generally, including property rights, are slow to change and, indeed, that lagged change accounts for many of the shortcomings of our natural resource policies. Nonetheless, property rights do change in response to changed social values. Following sections of this paper describe the larger social-legal framework that shapes the evolution of water rights in particular and that accounts for the remaining shortcomings of western U.S. water administration.

8.3 "Optimising" the Property Rights System

The property rights literature is full of ambiguities. Alan Randall speaks in favour of "unattenuated property rights" as an ideal (Randall 1985). Terms like "fully defined property rights" (Coase 1960; Anderson 1983), and "entitlements" (Calabresi and Melamed 1976) are seldom defined.[1] It is not clear in much of the literature whether more or less detail in the legal description of the property and related regulations lead to "more fully defined rights". From whose point of view is the right more complete? A lack of detail in the property right and related regulations allows latitude in the use of the asset, while detail narrows that discretion. The beneficiaries of the more restrictive right are usually a larger collective who stand to gain from the restrictions on the property owner's rights. That is, more detailed property rights may produce a collective good by restricting individual owner options. The same issue arises in the analysis of regulatory "takings" where collective values may be increased through improved environmental quality, safety, quiet, etc., even though individual owners' choices are restricted by the regulations. (Nelson 1996; Miceli and Segerson 1993; Epstein 1996). A common example of this trade-off is found in rural irrigation communities where individual

[1] Schmid (1987, p 8) implies a definition of property rights in stating that property rights are one way of disaggregating scarcity into individual entitlements; and one person's property rights constitute another person's limitation in the face of scarcity.

farmers desire maximum freedom to sell their water rights to the highest bidder while their communities push for restrictions on out-of-area water sales to preserve development potential.

Demsetz (1967) in his classic article envisions a process in which property rights evolve to a point where the marginal benefits from "internalizing another externality" equal the costs of that internalization. Demsetz presupposes that such rights will be fully enforced or, at a minimum, will lead to successful bargaining between the source and receptor of the externality. As the detail included in socially enforced property rights increases, private costs of litigation, bargaining and self-enforcement go down. Social costs of enforcement rise as more detailed effects must be monitored.[2] Other components of transaction costs such as search costs will rise because of the greater asset differentiation. Conceptually, we can thus define marginal benefit and cost curves as functions of the complexity of the property right definition.

These incremental costs and benefits will differ for different classes of assets and will shift with changes in the regulatory framework and liability rules that surround property rights. In some cases, the optimal complexity may be very high (e.g. for land that contains surface assets, mineral resources, public amenity values, water, etc.) and in other cases, it may be quite low (e.g. for rangelands with homogeneous attributes).

Optimization of institutional arrangements (a la Demsetz) cannot be carried out formally as we might use non-linear programming for the selection of a farm's cropping pattern. There are too many dimensions to the problem (economic, social, political) and some of the associated benefits and costs are difficult to quantify. Changes in institutional arrangements like the mix of property rights, regulations and liability should evolve from deliberations involving broad public input, hopefully informed by technical and scientific analyses. Some general guidelines can be stated.

The use of liability law to resolve externalities involves three complexities: (1) the establishment of "standing" in a particular court case, i.e. demonstrating sufficiently direct losses to make credible claims for compensation or injunction; (2) the construction of proof of causation of the damages; and (C) cost of court proceedings. Anticipated costs discourage many parties from proceeding with legal remedies. On balance, resort to tort law is likely to be restricted to relatively simple physical situations that involve large damages that justify the cost.

[2] Rapaczynski (1996) points out that "property rights can be only marginally enforced by the legal system... This means that most property-related arrangements ... must involve various self-enforced mechanisms ingrained in the incentive structure of spontaneous economic behaviour, so that moral virtue and the legal system are necessary to deal with only a thin layer of aberrational occurrence."

Regulations promulgated and enforced by public agencies may be the best way to reduce externalities where the physical system is too complex or the damaging externality too pervasive to make reliance on liability law feasible. Under these circumstances, reliance on property rights will also not be feasible. A prime example is the control of water quality where the system has many causes and the links between source and receptor are complex. Theoretically, a water right might specify both quantity and quality, but there is no physical possibility of guaranteeing water qualities tailored to each user. Regulation of emissions or the use of economic instruments (taxes and tradable permits) is then the answer. There are, of course, efficient and inefficient forms of regulation. With water quality regulation, it should be regulation of emissions and not the means of attaining those emissions.

One consideration should be the way in which regulations and tort issues affect the functioning of water markets. The volume of transactions must be sufficient to allow "price discovery", i.e. an estimation of price that will roughly equate buy and sell offers. Highly variegated property rights result in "thin" markets and high search costs.

8.4 The Evolution of Western U.S. Water Rights

The reader is referred to the papers of Miller (1995) and Scott and Georgina Coustatin (1995) for a complete historical account of the evolution of water law from Roman times through England to the United States. Suffice it here to state that the well-watered eastern U.S. adopted the English common law of riparian doctrine, while the later developing western U.S. developed its own priority (or appropriations) doctrine of water law. In early western irrigation and in the mining camps of California, it was necessary to divert water from streams, often transporting it through channels to distant fields or mining sites, contrary to riparian doctrine. It was also necessary to establish priorities of use so that those who got there first could be assured of continuing water supply. If an established use were abandoned, so that the water was no longer being "beneficially used", the water became subject to appropriation by others. Thus three main characteristics of western water law were established: (1) a priority system; requiring (2) diversion of the water from the stream; for (3) "beneficial use" (Bates et al. 1993). The need for diversion from the stream as a condition for a water right may have stemmed from fears of monopolization of the streamflow if it could be claimed for navigation or other instream purposes. Demonstration of "beneficial use" was intended to discourage "speculation", the bug-a-boo of all water administrators, even though beneficial use has been subject to a wide range of interpretation.[3]

[3] Casual history has it that the only use ever declared non-beneficial in Colorado was diversion of water by a farmer to drown prairie dogs.

In the early stages of development, this system served the major values of society quite well, putting water to use in productive pursuits that could support widespread resource exploitation. Given those needs, irrigated agriculture, ranching and mining flourished, but the environment often suffered as streams were dried up, riparian habitat was destroyed and water quality deteriorated.

In 1878, the Supreme Court of Colorado established the precedent that water rights were the personal property of the user and could be transferred to other uses, other locations and/or other owners. A right eventually came to be defined in terms of priority, the point and rate of diversion, the use of the water, the diversion dates and, for storage rights, the total volume allowed. Transfers of water rights to new uses and/or locations typically affect other parties on the stream, especially those who had become dependent on the return flows of the initial user.[4] Thus some type of social supervision of water appropriations and transfers was needed to assess negative externalities and, if necessary, to modify or prohibit the proposed transfer.[5]

The western states took different approaches to this supervisory task. Most states turned the task over to the agency responsible for water development and administration, usually the Office of the State Engineer. Such an office usually had expertise in hydrology and irrigation practice and could objectively determine the extent of third party impacts. The state of Colorado chose to use a court system within which those proposing water rights transfers had to defend the transfer against those who perceived themselves potentially injured. It seems likely that the court system has resulted in a litigious environment that has substantially increased transactions costs in comparison with other states' systems. The western systems have, nonetheless, permitted existing supplies to move to higher-valued uses through voluntary transactions that have benefited both buyer and seller, if not all parties.

---

[4] A water right is usually stated in terms of a rate of diversion from the stream. This diversion is partly consumed and part becomes return flow.

[5] Although reduction of externalities should be limited to those steps that generate more benefits than costs, the law insists on "no injury". However, a prospective buyer can contract with injured parties to drop their objections. While water law is concerned with negative externalities in the area of origin, it takes no cognizance of positive externalities below the new point of use.

8.5 The Need for Change in Western U.S. Water Administration

Water administration in some western states, while not exhibiting the rigidity sometimes claimed (Burness and Quirk 1979), has failed to take cognisance of a range of values of increasing social importance: water quality; riparian habitat; and a range of social and cultural values.[6] The social and community concerns include negative impacts on activities linked to irrigated agriculture; decreases in the local tax base resulting from reductions in irrigated acreage; and impacts on traditional cultures like the Hispanic communities centred on old acequia systems (irrigation ditches) of New Mexico and Arizona.[7]

However, progress has been made. The reach of some western administrative systems has been expanded to protect a greater range of "public values". Idaho has authorized the State Engineer to evaluate and, if necessary, to modify proposed water rights transfers to protect other water users, the "local public interest", the agricultural base of the community, fisheries, water quality, the local economic base and even the family farm (Administrator's Memo, Water Management Division, "Transfer Processing Policies and Procedures", available at Idaho Water Resource Board website; Costello and Kole 1985). Utah invokes the criteria of "no interference with irrigation, domestic, stock watering, power or mining; no unreasonable effect on public recreation or the environment" (available on the Holland and Hart website). Wyoming states that new appropriations or transfers should not impair the value of existing rights, no unacceptable economic loss to the community, and not be detrimental to the public welfare. New Mexico has the vaguest set of criteria: not to impair other water rights; not to be "contrary to the conservation of water" within the state; and not detrimental to the "public welfare".

While these criteria are wondrously vague and the degree of their enforcement unknown, they officially recognize the relevance of an increasing range of social values that are impacted by water administration. Colorado has the narrowest criterion for evaluating appropriations and transfers: no (water quantity) injury to other water rights owners. This does not include impacts on water quality that are obviously relevant. A legislative bill to allow (not require) the water courts to consider changes in water quality in water transfer cases were defeated in the 2003

---

[6] See Whitehead et al. 1995, for an interesting discussion of estimating non-use values.

[7] Concern with cultural impacts is rather new. In New Mexico, a District Court enjoined a transfer of water from an old acequia system to a new ski area, on the grounds that the traditional practices that held the community together would be disrupted (Sleeper vs New Mexico, 1987). Although reversed on appeal, this marked an advance in judicial recognition of cultural values. Also see Crawford (1988).

legislative session. Cities lobbied heavily against the bill while rural communities likely to lose water to the cities lobbied for the bill.

In spite of slow accommodation to changing values in the administration of western water, public opposition to large transfers of water out of the area of origin has been growing. In Colorado, legislation was introduced in 2003 to prohibit out-of-basin transfers of water rights on the grounds that important public values are being sacrificed in the areas of origin without compensation: secondary economic impacts related to decreased agricultural activity; reductions in the local tax base; forfeiture of future development possibilities; and disruption of cultural values. These secondary or "pecuniary" effects are what the public notices. Two questions should be answered before extreme restrictions are put in place: (1) which public values warrant "protection"? and (2) where "protection" of a public value is warranted, what form should that protection take?

The impacts noted above are called *externalities* of water transfers. Externalities usually involve physical impacts on third party utility or production, typically resulting in a loss of value (although positive externalities can occur. See footnote 5) that is uncompensated. "Pareto relevant" externalities are those portions of externalities for which the reduced damages (benefits of reduction) would exceed the cost of reducing the externality. Thus the extent of protection afforded public values can be informed by incremental benefit-cost tests.

Economists frequently distinguish two types of externalities: "real" (as used above) and "pecuniary" externalities. "Pecuniary externalities" of a transaction such as a water transfer are defined as changes in income levels of secondary economic activities that take place because of changes in market conditions, i.e. through demand and supply shifts and resultant price changes. Thus when irrigated land is dried up because of the sale of the water rights to a town, the local demand for farm inputs shifts downward, reducing employment and returns to capital among farm suppliers. The viability of farm product processors may also be negatively affected by the reduction of local supplies. Most economists conclude that these pecuniary externalities should be ignored, that this is "just the way the market works" to signal the need for the secondary resources to move to new uses.

This reasoning makes sense only if the unemployed resources will be quickly and costlessly absorbed elsewhere. When the sale of rural irrigation water to urban uses dries up productive acreage in areas that have long been economically depressed, there is likely to be long-term unemployment of human and other non-specialized (mobile) resources (Howe et al. 1990; Howe and Goemans 2003). Job search and moving costs are real, income losses that would have been avoided without the sale are temporary real costs, as are the costs of social disruption. Thus, some real economic efficiency costs are associated with "pecuniary externalities" and they may not be trivial in size. For example, with agricultural-to-urban water transfers, the erosion of the rural tax base reduces social services just when more are needed. In Crowley County, Colorado, the sale of water rights representing $1X10^6$ m^3 of irrigation water per year to the Denver area and the resultant drying-up of 16,000 ha of irrigated cropland resulted in a drop in the land tax base just at a time when expanded social services were needed. (Howe et al.1990). The only excuse for not including these costs in the assessment of water transfers is that they are difficult to estimate credibly and that the pecuniary externality concept has been abused in the past in exaggerating benefits of new water projects.

What can we conclude about protecting those public values impacted by direct, "pecuniary" and social externalities? First, steps to modify externalities should be subjected to marginal benefit-cost tests. "Corner solutions" (e.g. complete elimination) often recommended by environmental groups should not be accepted in most cases. Secondly, the real economic and social costs associated with "pecuniary" externalities should be taken into account insofar as they can be credibly estimated (for methods, see Howe and Goemans 2003). Finally, where the externalities are social and/or cultural and defy monetization, cost-effectiveness analyses of different protective steps might be undertaken to inform the political process.

8.6 Specific Steps

We have argued that several classes of economic and social values are not currently being adequately protected in the water transfer process e.g. water quality, ecosystem values and socio-cultural values. The use of water for the protection of such values should be considered "beneficial use" and any party (individuals, groups, private or public) seeking to achieve objectives through the voluntary purchase and application of water rights should be allowed to do so. "Beneficial use" doctrine is just regulation in a watery disguise.

Excessive water quality externalities can be mitigated in some settings through the purchase of water that has been applied to heavily saline lands, dedicating the water to instream dilution as recommended by Howe and Orr (1974) for control of salinity in the Colorado River. Water conservation trusts on agricultural lands are increasing in popularity in the ranching community. These market approaches have the advantage of forcing recognition of the *opportunity costs* involved in dedicating water to broad public objectives.

Finally, managing the speed with which water is transferred to new uses can be an important step since the social costs of change increase non-linearly with the speed of change. This is not a new idea since spreading the actual transfers over time is currently required in most western states to allow for revegetation of the formerly irrigated lands.

The arguments made above also imply that extra-market compensation to areas-of-origin are likely to be warranted in may cases. This, too, is not a new idea in water administration. The Colorado Water Conservancy District Act of 1937 requires that any project diverting water out of the Colorado River Basin in the State of Colorado must provide "compensatory storage" to provide a safety margin for the Colorado Basin in case of extreme water shortage. Naturally, added storage may not be the most efficient form of compensation. Compensation payments to units of general government are likely to be of greater value (Howe 2000). In a pioneering step, Aurora, Colorado has agreed to pay Otero County, the town of Rocky Ford and the Otero County School District, substantial sums for a number of years to compensate for the community impacts of drying up the Rocky Ford Ditch. It appears that such forms of consideration will be necessary to avoid increasing confrontation over rural-to-urban transfers.

Is there agreement that this is the way to go? While the observable drift of state water policies indicates that extra-market controls and compensation are needed to solve the increasingly complex issue of water allocation, some prominent voices decry the abandonment of the older, simpler forms of water markets. In a recent issue of the International Journal of Water Resources Development (2003), B. Delworth Gardner states:

> For more than half a century, in nearly all western states, the regulatory agencies...used impairment of other water rights as the primary criterion for approving or rejecting change applications to move water to higher-valued uses....the courts have considered whether additional criteria involving 'impacts on social welfare' are needed to evaluate transfers....

The critical question raised in this paper is whether existing state regulatory agencies can effectively implement a real 'social welfare' criterion to evaluate change applications. The conclusion is that they probably cannot... .

Anderson and Hill (Water Marketing-the Next Generation, 1997) state:

> Well defined property rights encourage people to search out margins of agreement, such as the farmer who discovers he can economize on water and profit from transferring some of his rights to a municipality. ...However, since some people do benefit from bureaucratic allocation of water, there are pressures for government control. This control has stood in the way of the evolution of property rights in water.

Haddad, in his book (2000) on the design of water markets for California provides some positive guidelines for the design of water allocation mechanisms appropriate to today's complexities:

> The world's current agenda with respect to markets has to do in part with how societies can control markets in ways that make the most of the good qualities but minimize their harm. Markets are being questioned on many fronts... . Markets are cropping up in unexpected places. ... Writers interested in markets no longer feel compelled to align themselves along the socialist-capitalist polarity... .This book is based on the idea that markets... have strengths and weaknesses that can be identified, described and sometimes measured. These qualities can then be compared with those of other resource-allocation mechanisms and a choice among them made....

Haddad's reasoned approach contrasts with the dogmatic faith in markets propounded by some. Today's complexities of values and goals require a major role for public agencies in water allocation and development. These agencies are, after all, instruments of representative government in which we must maintain faith.

8.7 References

Anderson TL, Hill PJ (eds) (1997) Water marketing-the next generation. Rowman and Littlefield, Lanham, MD

Anderson T (1983), Water crisis: ending the policy drought. The Johns Hopkins University Press, Baltimore, MD

Bates SF, Getches DH, MacDonnell LJ, Wilkinson CF (1993) Searching out the headwaters: change and rediscovery in western water policy. Island Press, Washington, DC

Binger B, Hoffman E (1988) Microeconomics with calculus. Scott, Foresman and Co, Glenview, IL

Brown FL, Ingram HM (1987) Water and poverty in the Southwest. University of Arizona Press, Tucson, AZ

Burness SH, Quirk JP (1979) Appropriative water rights and the efficient allocation of resources. American Economic Review 69: 25-37

Calabresi G, Melamed AD (1972) Property rules, liability rules, and inalienability: one view of the cathedral. Harvard Law Review 85: 1089-1128

Coase R (1960) The problem of social cost. Journal of Law & Economics 3: 1-44

Costello PD, Kole JP (1985) Commentary on swan falls resolution. Western Natural Resource Litigation Digest Summer: 11-18

Crawford S (1988) Mayordomo: chronicle of an acequia in northern New Mexico. University of New Mexico Press, Albuquerque, NM

Demsetz H (1967) Toward a theory of property rights. American Economic Review 57: 347-359

Epstein RA (1996) Takings: of private property and common. Conference paper presented at Economic and Constitutional Perspectives on Takings. American Enterprise Institute, Washington, DC, March 7

Gardner, DB (2003) Weakening water rights and efficient transfers. International Journal of Water Resources Development 19: 7-19

Greenley DA, Walsh RG, Young RA (1982) Economic benefits of improved water quality. Westview Press, Boulder, CO

Haddad BM (2000) Rivers of gold: designing markets to allocate water in California. Island Press, Washington, DC

Howe CW (2000) Protecting public values in a water market setting: improving water markets to increase economic efficiency and equity. University of Denver Water Law Review, 3: 357-372

Howe CW, Lazo JK, Weber KR (1990) The economic impacts of agriculture-to-urban water transfers in the area of origin: a case study of the Arkansas River Valley in Colorado. American Journal of Agric Econ December: 1200-1204

Howe CW, Orr DV, (1974) Economic incentives for salinity reduction and water conservation in the Colorado River Basin. In: Flack E, Howe CW (eds) Salinity in Water Resources. Merriman Publishing Co, Boulder, CO

Miceli TJ, Segerson K (1993) Regulating agricultural groundwater contamination: a comment. .Journal of Env Econ & Mgt 25: 196-200

Milgrom P, Roberts J (1992) Economics, organization and management. Prentice-Hall, Englewood Cliffs, NJ

Miller KA (1995) Property and water: a historical analysis. National Centre for Atmospheric Research, Boulder, CO

Nelson H (1996) The takings provision of the contract with America. A paper presented at the annual ASSA/Association of Environmental and Resource Economists meeting, San Francisco, CA Jan. 5-7

Randall A (1985) Resource economics: an economic approach to natural resource and environmental policy. John Wiley & Sons, NY

Rapaczynski A (1996) The roles of the state and the market in establishing property rights. Journal of Economic Perspectives, 10: 87-103

Schlager E, Ostrom E (1993) Property rights regimes and coastal fisheries: an empirical analysis. In: Anderson TL, Simmons RT (eds) The political economy of customs and culture: informal solutions to the commons problem. Rowman and Littlefield, Lanham, MD

Schmid A (1987) Property, power and public choice: an inquiry into law & economics. Praeger Publishers, NY

Scott A, Coustatin G (1995) The evolution of water rights. Natural Resources Journal, 35: 821-980

Varian HR (1978) Microeconomic analysis. W. W. Norton & Co, NY

Whitehead JC, Blomquist GC, Hoban TJ, Clifford WB (1995) Assessing the validity and reliability of contingent values: a comparison of on-site users, and non-users. Jour of Env Econ & Mgt 29: 238-251

Finding a Modern Role for the Prior Appropriation Doctrine in the American West

Ray Huffaker

9.1 Introduction

U.S. Interior Secretary Gale Norton called western water supply conflicts "one of the greatest challenges facing the nation in the coming decades" (Hanson 2003). Water supplies fall short of satisfying all of the demands placed upon them, even absent drought. In one example, the federal government's decision to shut down irrigation during the 2001 season in Oregon's Klamath Basin to preserve fish species listed under the Endangered Species Act sparked great consternation on the part of irrigators and agricultural communities (Millstein 2002a). Alternatively, the federal government's decision to provide water for irrigation during the 2002 season, and its possible role in causing the massive die-off of 20,000 to 30,000 adult migrating salmon in the Klamath River due to inadequate river levels, greatly alarmed fishing and environmental interests (Millstein 2002b).

Western water supply conflicts have their roots in the set of laws and institutions governing western water use, and the manner in which these laws have been implemented. The prior appropriation doctrine has provided the foundation of this law for over one hundred years. Historically, the doctrine promoted the rapid development of the American West by securing water to irrigated agriculture when that sector constituted the major portion of economic activity and social welfare. Indeed, the doctrine has permitted irrigated agriculture to appropriate the vast bulk of dependable river flows in the west (80-90 percent). However, in modern times, the prior appropriation doctrine has proven to be a liability in reallocating water to emerging non-appropriative uses whose contribution to social welfare increases as irrigated agriculture's historic contribution declines.

This chapter will: (1) discuss the central tenets of the prior appropriation doctrine; (2) investigate how intrinsic flaws in these tenets, and the way in which they are applied, render the doctrine inflexible for satisfying the full range of modern water needs; and (3) consider the extent to which the prior appropriation doctrine can be modified for increased flexibility, and can work compatibly with emerging alternative doctrines protecting non-appropriative uses, principally the public choice doctrine.

9.2 The Prior Appropriation Doctrine

The prior appropriation doctrine has its roots in the rule of capture characterizing western mining law: "first in time, first in right." (Wilkinson 1992). A person obtains a user's (usufructuary) right to the quantity of publicly-owned water first diverted to a beneficial use on a fixed tract of appurtenant land. The quantity of water under the right is called the "water duty." In the case of irrigation, it is measured by the amount of water needed to irrigate an average mix of crops on the appurtenant land with the irrigation technology prevailing when the water right was granted. The priority of the right reaches back to the date of the first diversion. When water supplies fall short, the earliest ("senior") appropriators receive their full entitlements until supply is exhausted. The rights of later ("junior") appropriators are either curtailed or completely cut off. Senior appropriators seeking to expand their diversionary rights beyond the water duty must execute a new water right having the most junior priority ("no expansion of use"). Water that is not beneficially used is forfeited and available for re-appropriation by another user ("use it or lose it").

Elwood Mead, an early pioneer in western water policy, described the protection that prior appropriative water rights theoretically offer senior and junior water users as follows:

"As scarcity of water led to the adoption of the doctrine of priority, the two led to the necessity of defining the quantity of water to which an appropriator should be entitled. While the early appropriators were entitled to protection in their use of water, the later comers had an equal claim to protection from enlargement of those uses. The first appropriator had the first right, but he had not the right to take all the water he might want at any future time. His right must, in justice to others, be defined as to quantity as well as to time. In theory, beneficial use has been made the measure of a right. That is, each appropriator has a right as against a subsequent appropriator to a continued use of whatever quantity of water had been put to a beneficial use at the time of the acquirement of the subsequent right." (Mead 1903 quoted in Gould 1988).

The protection described in this quote depends on the maintenance of complicated interrelationships created among water users due to the flowing nature of water. Crops generally do not consume all of the water that is diverted from a stream. The unconsumed portion can return to the stream as surface runoff or as underground spring flow after deep percolation to an underlying aquifer ("return flows"). Alternatively, it may escape by the same means to a second water course, or be irretrievably lost due to, for example, surface evaporation ("escape flows"). Return flows and escape flows (that are not irretrievably lost), along with natural flows, supply a significant portion of the water required to satisfy water rights in the west. One study estimated that return/escape flows constitute almost half of all water diverted in the west (Pulver 1988). Consequently, as discussed below, water rights depending on return/escape flows may be impaired if some users are allowed to make changes in water use that reduce the level of these flows.

9.3 Flaws in Principle and in Application

Most would agree that the prior appropriation doctrine played an important role in promoting the rapid economic development of the West. Some contend that the doctrine continues to provide important benefits in allocating water by: (1) encouraging the full utilization of water where needed without tying its use to the ownership of riparian land; (2) discouraging wasteful use with the beneficial-use and use-it-or-lose-it requirements; and (3) stimulating economic investment in water resources by providing for secure water rights, allocating water in a fair way, and being relatively easy to administer (Gopalakrishnan 1973; Gould 1988).

This section investigates how the prior appropriation doctrine has ceased to provide these traditional benefits on a consistent basis. One major reason is that the nineteenth century doctrine has not adapted well to change. It functioned best when circumstances remained as they were when appropriative rights were granted, but, of course, circumstances inevitably have changed. In particular, the widespread and gradual adoption of improved agricultural water application technologies has been eroding the security of prior appropriative rights and increasing the pressure on already scarce water supplies. In addition, the doctrine has proven to be too inflexible to provide a reliable water supply for emerging instream flow uses in the modern era. A second major reason that the doctrine has ceased to provide its traditional benefits is a problem of application, in particular, a lack of enforcement. This section discusses how states have not been vigilant in policing illegal appropriations and enlargements of water rights.

This section contends that, in the face of responding to these challenges, the doctrine's principles devolve into a tangled web of inconsistencies that conceals allocation problems and frustrates the formulation of rational water allocation policies. As a result, water rights holders are looking to state legislatures, agencies, and courts to provide protection beyond that offered by prior appropriation doctrine.

9.3.1 Response to the Adoption of Improved Irrigation Technologies

The west has witnessed gradual and widespread increases in the efficiency with which water is applied in irrigation. Irrigation scientists calculate on-farm irrigation efficiency (IE) as the ratio of the water consumptively used by crops (C) to the total water diverted for irrigation (D), i.e., $IE = C/D$ (Wade 1986). Nineteenth century irrigation technologies and management skills resulted in low irrigation efficiencies in the neighbourhood of 25 percent. For example, a consumptive use requirement of 2 units of water would require the farmer to divert 8 (= 2/0.25) units to the field. The remaining 6 (= 8-2) units of water would constitute return/escape flows supplying a portion of the water diverted by other appropriators. Farmers have improved irrigation efficiency by applying water in a more timely and uniform manner across their fields. Conversions of traditional gravity flow irrigation systems to sprinkler technology have increased efficiency to 75 percent and above.

Farmers increasing irrigation efficiency can have deleterious impacts on the return/escape flows relied upon by other water users if the farmers are not restricted to the level of consumptive use characterizing their original water rights. We use the streamflow diagrams in Figure 9 to demonstrate how such impairment can occur in a hypothetical river basin. Figure 9(a) depicts status quo circumstances in which an average weather year provides 16 units of stream inflow to the basin during the irrigation season. There are two farms, A and B, each of which irrigates with a traditional gravity-flow technology at 25 percent efficiency that existed at the time that their water rights were granted. Farm A has a right to divert all 16 units of inflow that is senior to Farm B's right to divert 12 units. Farm A's crops consume 16(0.25) = 4 units (encircled quantity), and the difference between the diversion and consumptive use is 16-4 = 12 units. These 12 units return to the stream, and supply the quantity of water needed to fill Farm B's water right. Farm B's consumptive use is 12(0.25) = 3 units. The ensuing return flow is 12-3 = 9 units, which constitutes the stream flow at the bottom of the basin.

Now consider the impact on Farm B's water right if Farm A alters the status quo by adopting a modern irrigation technology that increases efficiency to 75 percent. Farm B's irrigation efficiency remains at 25 percent. We look at two scenarios. In the first [Figure 9(b)], Farm A is allowed to expand its consumptive water use at the expense of Farm B and instream flow levels at the bottom of the basin. Farm A might expand consumptive water use by making other changes away from the status quo, such as growing more water intensive crops or increasing acreage beyond that appurtenant to the original water right (i.e., water spreading). In particular, Farm A's crops now consume 16(0.75) = 12 units, and return flow is reduced by 12-4 = 8 units compared with the status quo. As a result, Farm B's water right is curtailed by these same 8 units (i.e., it only can divert 4 instead of the original 12 units), and consumptive water use by B's crops is reduced by a third (i.e., from 3 units to 1 unit). In sum, the 8 units of increased consumptive use on Farm A is funded by reductions of 2 units of consumptive use on Farm B and 6 units of reduced stream flow at the bottom of the basin.

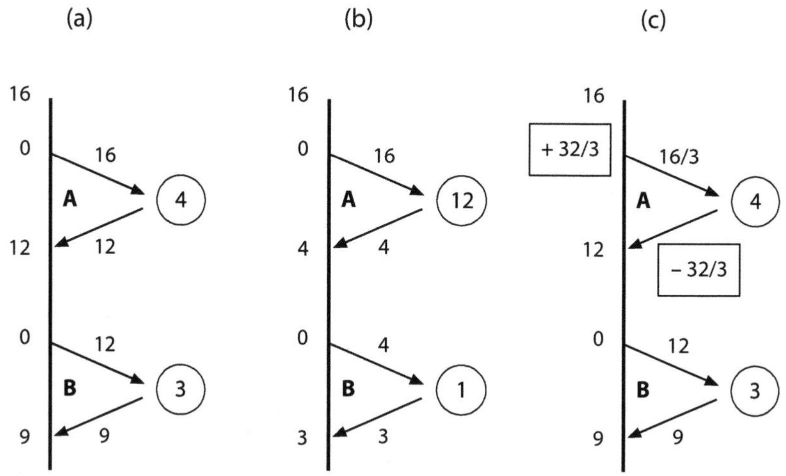

Fig. 9.1 Increases in on-farm irrigation efficiency: (a) status quo with 25 percent efficiency, (b) farm A increases efficiency to 75 percent and is allowed to increase consumptive use, (c) farm A increases efficiency to 75 percent but is held to status quo level of consumptive use.

In the second scenario [Figure 9(c)], Farm B appropriative right is unimpaired because Farm A is held to its status quo level of consumptive use. In particular, increased efficiency means that Farm A can reduce the diversion required to generate 4 units of consumptive use by a third (i.e., 16 units to 16/3 = C/IE = 4/0.75 units). This increases stream flow at Farm A's point of diversion by 32/3 units [boxed quantity], but reduces return flow by the same amount [other boxed quantity]. The result of this offsetting change is to preserve Farm B's status quo in Figure 9(a). There are 12 units of water available to provide for Farm B's full appropriative right, and 9 units of water remain at the bottom of the basin.

In the real world, policy makers have not preserved the status quo for other appropriators and instream uses because they have allowed efficiency-increasing farms to spread water over additional acreage. One example is the Columbia Basin Project in Washington state. It is an irrigation project run by the United States Bureau of Reclamation that initially defined lands adaptable for irrigation based on the flood irrigation technology prevailing in the 1940's. The water duty for these lands was established in excess of 5 acre-feet of diversion per acre. The large water duty, along with shifts to centre-pivot irrigation, now permits irrigation of many acres previously judged to be non-irrigable. Water spreading occurs without challenge from state or federal institutions, thereby politically validating the enlargement of the original water rights to include more land and greater water consumption.

Another example is provided by upper Snake River Plain aquifer. Traditional flood-irrigation practices were highly inefficient, which allowed a large quantity of unconsumed irrigation diversions to percolate into the aquifer. This substantially increased historic aquifer levels and augmented downstream flows in the Thousand Springs reach of the Snake River. A switch to more efficient irrigation practices has caused a precipitous decline in aquifer levels and subsequent recharge of the lower portion of the Snake River. Appropriators diverting water in this stretch of the river find their rights impaired (Johnson et al. 1999).

Why water administrators have not protected other appropriators and instream-flow rights from impairment by restricting efficiency-improving appropriators to their traditional level of consumptive use, as in Figure 9(c)? A major reason is that the basic principles of the prior appropriation doctrine do not compel such a solution, and, in fact, create contradictions that seem to justify allowing impairment.

Contradiction arises because the water duty, defined in terms of diversionary quantities, fails to account for the impact of consumptive use on return flow. Consequently, efficiency-improving irrigators can claim ownership over the portion of the diversionary quantity no longer required to meet the original level of consumptive use as part of the original water duty [i.e., the 32/3 units of water in Figure 9(c)], and thus legal entitlement to spread it over additional crops. Otherwise, they can contend that the unused portion will be forfeited subject to the "use-it-or-lose-it" principle, which constitutes a public appropriation of part of their original water duties. Alternatively, opponents can contend that efficiency-improving irrigators have enlarged their water use illegally by spreading water over land non-appurtenant to their original right. This impairs the water rights of other appropriators contrary to the intent of the prior appropriation doctrine.

Policy-makers in several western states have attempted to resolve the contradiction by adopting two convenient, but unjustifiable, fictions. First, they presume that the unused portion of the water duty after an increase in irrigation efficiency constitutes "conserved" water that can be spread without impairing other water rights. For example, agricultural water conservation statutes in Oregon and Washington define conserved water as the reduction in required diversions before and after an increase in on-farm irrigation efficiency [i.e., 32/3 units in Figure 9(c)] without regard for impacts on consumptive water use and return flows. They further award various portions of the "conserved" water to efficiency-improving irrigators [see, Oregon Revised Statutes §537.455(1) and Revised Code of Washington §90.03.380(1)]. Federal policy-makers also make this presumption. For example, Secretary Norton's "Water 2025" initiative designed to resolve western water-supply crises calls for making irrigation more efficient (Hanson 2003).

Unfortunately, as demonstrated in Figure 9(c), the reduction in the diversion required to maintain traditional levels of consumptive use after an increase in irrigation efficiency is offset by a reduction in the irrigator's return flow. As a result, stream flow below the efficiency-improving irrigator's point of return flow remains unchanged. No new water is created. However, if the law misguidedly allows the efficiency-improving irrigator to spread this water anyway, then streamflow will decrease below the point of return flow in the amount of the increase in consumptive use, and the rights of downstream appropriators and instream-flow uses will be impaired as demonstrated in Figure 9(b).

A second fiction that states have adopted to allow efficiency-improving irrigators to spread water at the expense of other water rights is that irrigation return flow constitutes "waste" water in violation of the beneficial-use requirement. Consider, for example, the case of *Estate of Steed v. New Escalante Irrigation Co.* [846 P.2d 1223 (Utah 1992)]. The New Escalante Irrigation Co. diverted water from the Escalante River in Utah and applied it to crops with an inefficient irrigation technology. Water unconsumed by crops escaped to Alvey Wash where the Steed estate had a vested appropriative right to divert it for irrigation. New Escalante subsequently converted to a more efficient irrigation technology that substantially reduced escape flows to Alvey Wash, and thereby impaired the Steed estate's vested water right. The court refused to protect the Steed estate's vested water right after concluding that it drew water from a previously wasteful irrigation practice. The court failed to realize that escape flows, resulting from the inefficient irrigation practices existing at the time many appropriative rights were granted, are a legitimate means of supplying other water rights. Water is not wasted, it is simply used by another appropriator. As a result, New Escalante was allowed to enlarge its water use contrary to the intent of the prior appropriation doctrine, and in the process impair the Steed estate's water right.

To summarize, the adoption of improved technologies for applying irrigation water to crops brings several benefits to the efficiency-improving irrigator, including increased crop production and lower irrigation costs. Unfortunately, these benefits may come at the expense of other water-right holders and instream uses if efficiency-improving irrigators are allowed to increase their traditional levels of consumptive water use. The result is increased pressure on already over appropriated western water systems. Inconsistency arises as the prior-appropriative principles of beneficial use, use-it-or-lose-it, and the measurement of the water duty in terms of diversionary quantities appear to justify allowing efficiency-improving irrigators to spread water; while principles restricting diversion to appurtenant land and prohibiting enlargement of use appear to ban spreading. The adoption of unjustifiable fictions by federal and state water agencies does not resolve this inconsistency. It is like ignoring a disease as it evolves into an epidemic.

9.3.2 Response to Emerging Water Demands for Instream Flow Uses

The prior appropriation doctrine has not adapted well to changing social demands for water. By definition, instream uses of water do not meet the doctrine's traditional requirement that water be 'diverted to a beneficial use.' Due to this lack of recognition, instream uses were locked out as irrigated agriculture appropriated the vast bulk of dependable river flows in the west (80-90 percent) in the late nineteenth and twentieth centuries. State appropriation systems continue to restrict the dedication of appropriated water to instream use (Huffman 1983). For example, water rights for instream use generally are established either by states or state-sanctioned water trusts, and are held by the states in trust. They have junior priority to appropriative rights established before them (Hamilton and Huffaker 2003).

As a result, irrigated agriculture maintains prior rights regardless of how little water remains for instream uses, or how much more valuable these uses might be at the margin. The evidence is that the potential gains from marginally shifting water to instream uses may be large. For example, two economic analyses concluded that shifting water to hydropower production would generate benefits ten times greater than lost farm income (Hamilton et al. 1989), and two times greater if the water were shifted primarily to provide flows for migrating fish species (Hamilton and Whittlesey 1992).

Markets have been long recommended as a means of transferring water to more profitable uses within the framework of prior appropriation (Milliman 1959). However, they have not developed substantially for several reasons. One reason is that, similar to the impact of improved irrigation technology, the transfer of appropriative rights can compromise the security of other rights by changing the timing, quantity, and quality of return/escape flows. States protect appropriators from such impairment by imposing moderate to severe limits on water transfers (Gould 1988). For example, Nebraska traditionally has banned the transfer of appropriated water to non-agricultural uses. Wyoming traditionally has severely limited transfers by placing a heavy burden on the transfer proponent to prove lack of injury to other appropriative rights with 'clear and convincing' evidence (Huffaker and Whittlesey 2000).

Another reason that water marketing has not developed more substantially is that it provokes heated political opposition from rural communities fearing that transferring water out of agriculture will have disastrous impacts on regional employment, businesses, and tax bases (see e.g., Hymon 2002). Consequently, although many farmers may be willing sellers, local agribusiness and community interests actively work against marketing policies. For this reason, a voluntary buyout programme, in which the federal government would purchase farm interests in water and/or land for use in the Klamath Basin's National Wildlife Refuges, was dropped although 24 farm families controlling 30,000 acres were offering to participate (ONRC 2001).

9.3.3 Lack of Enforcement of Prior Appropriation Principles

The security of water rights sanctioned under state prior appropriation systems is imperilled by noncompliant water use that state water agencies allow to continue unabated, and state legislatures attempt to legitimise by statute. For example, environmental groups sued the state of Washington to force the state's Department of Ecology to ban noncompliant groundwater withdrawals (i.e., groundwater pumped beyond levels specified in water rights) because of their detrimental impacts on instream flow (Spokane Spokesman Review 1999). In another example, a 1993 survey found that over 500 water users (mostly irrigators) were taking water without a valid right in Whatcom County, Washington. The 1997 Washington Legislature attempted to legitimise these invalid uses by passing a bill that was subsequently vetoed by the Governor (Benson 1998).

The efforts of state water agencies to identify and stop noncompliant use are often plagued by inadequate budgets (Spokane Spokesman Review 1999), and by inconsistent enforcement policies set out by state legislatures and courts. For example, the Idaho Supreme Court held that the state water agency has a mandatory duty to curtail noncompliant water use to protect senior appropriators. Alternatively, a Washington State Court ruled that the state water agency (which ideally has the expertise to untangle complex water issues) is unauthorized to evaluate the priorities of water rights that have not undergone a formal basin-wide adjudication procedure, or to issue regulatory orders to persons violating those rights (Rettkowski 1993). A leading water attorney in the state analysed the impact of this case on the State's prior appropriation system as follows: "Since most of the state's waters remain unadjudicated, the most obvious effect of [Rettkowski v. Department of Ecology] is that for most water users, priority—the keystone of Western water law—is now meaningless..." (Dufford 1994)

More recently, a Washington State Court mediated an agreement ending litigation between the Washington Department of Ecology and Columbia River irrigators that appears to remove the irrigator's rights outside of the normal prior-appropriative framework (Voluntary Settlement Agreement 2002). Irrigators sued to force the state to issue seven water rights. The settlement outlines two ways in which irrigators can be issued rights that the state guarantees will not be interrupted when flows go below instream flow levels set in 1980. Irrigators can either pay $10 per year per acre-foot of uninterrupted water (so that the state can buy replacement water), or install state-of-the-art irrigation devices. Water "conserved" from the increased efficiency would be transferred to the state. The agreement does not consider where replacement water can be purchased for as low as $10 an acre foot in an already over appropriated river basin. It also does not acknowledge the fallacy of equating increased on-farm irrigation efficiency with agricultural water conservation in the return-flow hydrology characterizing the Columbia and Snake River basins [Figure 9(c)]. The settlement effectively grants these irrigators a new species of water right outside the realm of the prior appropriation doctrine. Contrary to the principle of 'first in time first in right,' senior water right holders with 'interruptible' rights would be curtailed by these 'uninterruptible' junior rights in water short years.

9.4 A Modern Role for the Prior Appropriation Doctrine

The modern-day erosion of the prior-appropriation doctrine's past virtues undermines its past status as the pre-eminent water allocation mechanism in the West. Wilkinson (1992) has characterized the doctrine as one of several outmoded western resource laws ('lords of yesterday') that have failed to keep pace with societal change, and thus constitute "a government of living by the dead" (p 25). He punctuated his frustration by composing a tongue-in-cheek obituary entitled, "West's Grand Old Water Doctrine Dies" (Wilkinson 1991). There is, at a minimum, the need to operate with a more flexible version of the prior appropriation doctrine in conjunction with parallel allocation doctrines better equipped to distribute water to a number of meritorious competing public uses.

9.4.1 Increasing the Flexibility of the Prior Appropriation Doctrine

Several ways have been recommended to make the prior appropriation doctrine more responsive to modern water allocation needs while offering some degree of protection for traditional water rights. One recommendation is to restrict water transfers to the quantity of water consumptively used under the right. This can mitigate some, but not all, of the quantity-related impairments to water rights not involved in the transaction (Anderson and Johnson 1986). Moreover, such a quantity restriction is not designed to solve problems related to the timing or quality of return flows (Gould 1988). Due to these limitations, economists have recommended various types of specialized transfers designed to limit the extent and duration of transfer-related impairments. These include 'trial transfers' (transfers that can be modified or revoked if actual impairment results), 'one-time-temporary transfers' (transfers whose short terms make injuries of short duration), and 'contingent transfers' (transfers that occur intermittently and are triggered only by some predetermined contingency).

A second recommendation for increasing the flexibility of the prior appropriation doctrine is to lower the transaction costs of transferring water. For example, the administrative mechanisms supervising transactions can be streamlined by formulating rules to narrow the range of technical issues addressed in each transfer hearing (Gould 1988; Colby 1990), shifting the burden of proof to other appropriators claiming impairment (Gould 1988), and crafting expedited review mechanisms for the specialized water transfers discussed above (Gould 1988).

A third recommendation is for states to remove outright restrictions, and less transparent legal disincentives, on the transfer of water outside of agriculture.

The last recommendation mentioned here is to remove the 'use it or lose it' requirement from prior appropriative rights. The logic behind this traditional restriction is that inappropriate water is wasted. However, this logic no longer holds in modern times when the ecological value of instream flow is recognized. Irrigators have the adverse incentive to continue employing water in a sub-economic use rather than to transfer it to a more valued instream use, since transferred water might be lost to the use-it-or-lose-it requirement (Economist 2001).

9.4.2 Parallel Water Allocation Doctrines

The public trust doctrine is a parallel water allocation doctrine with historical roots deeper than those of prior appropriation. The doctrine is an evolution of common-law principles dating from medieval England that recognizes the government's obligation to manage some types of natural resources in trust for the public benefit (Stevens 1980). For example, in medieval England, public rights protected by the trust were navigation, transportation, and fishing on lands along seashores and rivers. These rights accompanied any grant of the lands by the King into private ownership. In the USA, the US Supreme Court and State courts determine the resources covered under the trust and the range of public activities to be protected. For example, the US Supreme Court has held that navigable freshwater bodies and tidelands are covered by the public trust. Individual states have extended public trust obligations to rural parklands; wetlands associated with navigable water bodies; non-navigable tributaries; and waters usable for fish and wildlife habitat and recreational purposes. Similar to English common law, these public trust obligations accompany the transfer of trust resources into private ownership or use. Consequently, federal and state governments are empowered by the trust to condition appropriative water rights to the extent required to protect public resources and uses protected under the trust. For example, government entities could reject applications for new appropriations that would harm trust values, and continually supervise and regulate existing rights to ensure compatibility.

Federal and state governments also are empowered to condition appropriative water rights by regulatory programs designed to preserve public resources. For example, the Endangered Species Act authorizes lead federal agencies to formulate a recovery plan for listed species, and demands compliance from federal and state agencies (section 7), and private citizens (section 9). The federal government has exercised this power to shut off irrigation water supplied by federal projects, or conveyed by federal ditches, to Washington State's Methow Valley and Oregon's Klamath basin.

9.5 Concluding Remarks

Prior appropriation is a nineteenth century doctrine that functioned best for the economic and technological circumstances of that era. Irrigated agriculture was able to appropriate the vast bulk of water because it constituted the major portion of economic activity, and instream uses were not yet recognized. Water application efficiencies in agriculture were low which meant that a relatively small fraction of diverted water was consumed by crops, and a relatively large fraction returned to the hydrologic system for use by downstream appropriators.

The modern era has witnessed the emergence of instream uses whose contribution to social welfare is rising as irrigated agriculture's historic contribution declines. Competition for water between agricultural and instream ecological and hydropower uses has stressed the allocative abilities of prior appropriation to the crisis point, as illustrated by Oregon's Klamath Basin. Such stress might have been relieved before the crisis point if the prior appropriation doctrine were amenable to water marketing on a wide geographic scale. However, the doctrine's inflexibility in transferring water to meet instream needs has resulted in its occasional subjugation to federal laws (e.g., the Endangered Species Act) designed to protect these uses.

The modern era also has witnessed vast improvement in on-farm irrigation efficiency that has increased agricultural productivity and reduced irrigation costs. However, it also has tended to increase consumptive water use in agriculture, reduce irrigation return flows, and, consequently, further stress over appropriated water supplies. The principles of prior appropriation are not designed to identify or remedy the impairment in downstream water rights deriving from increased irrigation efficiency upstream. As a result, the protection that the doctrine traditionally has offered to appropriative water rights is eroding, and rights holders are looking to state legislatures, agencies, and courts to provide protection in other ways.

To the extent that the prior appropriation doctrine continues to play a modern role in western water allocation, it should be made more flexible in allocating water to socially-desired instream uses, and be operated in conjunction with parallel allocation doctrines better suited to protect these uses. The public ownership of water, the public trust doctrine protecting trust resources and values, and federal environmental statutes provide ample authority for public officials to condition appropriative water rights as required to protect instream uses.

9.6 References

Anderson T, Johnson R (1986) The problem of instream flows. Economic Inquiry 24:535-554

Benson R (1998) Maintaining the status quo: protecting established water uses in the Pacific Northwest despite the role of prior appropriation. Research Publication PO97-1, Northwest Water Law & Policy Project

Colby B (1990) Transactions costs and efficiency in western water allocation. American Journal of Agricultural Economics 72:1184-1192

Dufford W (1994) Water law after Sinking Creek. In: University of Washington (ed) Proceedings of the Sinking Creek decision: water rights in the 21st century, pp A2-A12

Economist (2001) You say potato, I say electricity. July 12, 2001, p 28

Gopalakrishnan C (1973) The doctrine of prior appropriation and its impact on water development: a critical survey. American Journal of Economics & Society 32:61-72

Gould G (1988) Water rights transfers and third-party effects. Land and Water Law Review 23:1-41

Hamilton J, Whittlesey N (1992) Contingent water markets for salmon recovery. (Unpublished working paper, University of Idaho, Department of Agricultural Economics)

Hamilton J, Whittlesey N, Halverson P (1989) Interruptible water markets in the Pacific Northwest. American Journal of Agricultural Economics 71:63-73

Hamilton J, Huffaker R (2003) Conflict in US Irrigation. In: Lascano R, Sojka R (eds) Irrigation of Agricultural Crops Monograph. USDA-ARS, in press

Hanson D (2003) Feds to take on Western water conflicts. The Spokesman-Review. May 3, 2003

Huffaker R, Whittlesey N (2000) The role of prior appropriation in allocating water resources into the 21st century. International Journal of Water Resources Development 16:265-273

Huffman J (1983) Instream water use: public and private alternatives. In: Anderson T (ed) Water rights. Pacific Institute, San Francisco, pp 249-282

Hymon S (2002) Klamath's water is better used downriver, study finds. The Los Angeles Times, November 5, 2002

Johnson G, Sullivan G, Cosgrove D, Schmidt R (1999) Recharge of the Snake River Plain aquifer: transitioning from incidental to managed. Journal of the American Water Resources Association 35:123-131

Milliman J (1959) Water law and private decision making: a critique. Journal of Law and Economics 2:41-63

Millstein M (2002a) Top official pledges study of fish die-off. The Oregonian. October 3, 2002

Millstein M (2002b) Klamath diagnosis is a warning for future. The Oregonian. December 19, 2002

Oregon Natural Resources Council (ONRC) (2001) Farmers and conservationists agree on solution for Klamath Basin water crisis, in http://www.org.onrc.org, June 15, 2001

Pulver R (1988) Liability rules as a solution to the problem of waste in western water law: an economic analysis. California Law Review 76:671, citing US Bureau of Reclamation 1973, Shut Off the Water—The Root Zone is Full: A Study of Irrigation Water Use

Rettkowski V, (1993) Department of Ecology, 858 P 2d 232

Stevens J (1980) The public trust: a sovereign's ancient prerogative becomes the people's environmental right. University of California Davis Law Review 14:195

The Spokane Spokesman Review (1999) Water suit against agency headed to court. December 4, 1999

Voluntary Settlement Agreement (2002) Benton County Superior Court Cases No. 09-2-82852-7 and 97-2-01041-9. November 20, 2002
Wade J (1982) Efficiency and optimization in irrigation analysis. In: Whittlesey N (ed) Energy and water management in western irrigated agriculture. Westview Press, Boulder, pp 73-100
Wilkinson C (1992) Crossing the next meridian. Island Press, Washington, DC
Wilkinson C (1991) West's grand old water doctrine dies. High Country News 23:1-14

Index

Q

R

S

T

X

Y

Z

MIX
Papier aus verantwortungsvollen Quellen
Paper from responsible sources
FSC® C105338

If you have any concerns about our products,
you can contact us on
ProductSafety@springernature.com

In case Publisher is established outside the EU,
the EU authorized representative is:
Springer Nature Customer Service Center GmbH
Europaplatz 3, 69115 Heidelberg, Germany

Printed by Libri Plureos GmbH
in Hamburg, Germany